T0205496

Underwater Acoustic Channel

Junying Hui · Xueli Sheng

Underwater Acoustic Channel

Harbin Engineering University Press

Springer

Junying Hui
College of Underwater Acoustic
Engineering
Harbin Engineering University
Harbin, Heilongjiang, China

Xueli Sheng
College of Underwater Acoustic
Engineering
Harbin Engineering University
Harbin, Heilongjiang, China

ISBN 978-981-19-0776-0   ISBN 978-981-19-0774-6 (eBook)
https://doi.org/10.1007/978-981-19-0774-6

Jointly published with Harbin Engineering University Press, Harbin, China
The print edition is not for sale in China (Mainland). Customers from China (Mainland) please order the print book from: Harbin Engineering University Press.

This Springer imprint is published by the registered company Springer Nature Singapore Pte Ltd.
The registered company address is: 152 Beach Road, #21-01/04 Gateway East, Singapore 189721, Singapore

Preface

The ocean is the resources treasure of humans. Also, it is the battlefield of war. With the continuous improvement of science and technology, the informationization and modernization of marine research, and the rational exploitation of marine resources are attracting more and more attention. The sound waves travel farther than any other physical field in the ocean, and they have the unique advantage of becoming a powerful tool for detecting and studying the ocean. Thus, the underwater acoustic technique has been developed. With the exploration and development of marine energy and mineral resources, as well as the traction of military demand, underwater acoustic technique has become an essential acoustic application subject, and underwater acoustic technique in China has also made great progress in recent years.

Underwater acoustic equipment must be matched with the environment to achieve the goal of designing optimally. However, in general, electronic engineers have little understanding of the effects of the marine environment on sonar systems, and underwater acoustic physicists, despite their extensive knowledge of underwater acoustic channels, do not have a deep understanding of engineering design needs. This book will build a "bridge" between them and try to discuss the limitations and impacts of ocean acoustic channels on sonar systems in mathematics and language that engineers can understand, especially from the perspective of sonar signal processing to study the ocean. Therefore, the ocean is regarded as an underwater "acoustic channel".

In terms of communication theory, the ocean between source and receiver hydrophone is regarded as a linear and complex stochastic filter with time-varying and space-varying. *Underwater Acoustic Channel* is to study the structure and characteristics of this filter and the influence on sonar signal processing to possibly match the sonar signal processor with the marine environment. If people do not understand the complexity of underwater acoustic channels, they will not achieve the design goal. If the complexity is overestimated, the design cost will be too high. Actually, the environmental conditions of underwater acoustic equipment vary significantly for different applications. One can regard the underwater acoustic channel, for most applications, as "slow time-varying, space-varying coherent multipath channel", which is the basic view of this book.

There are nine chapters in this book. Chapter 1 briefly reviews the basics of acoustics and sonar equations. In Chap. 2, the characteristics of ocean channels with only propagation loss are discussed. From Chaps. 3–6, the time-varying and space-varying characteristics of underwater acoustic channel are analyzed. These four chapters present the perspective that the underwater acoustic channel is regarded as "slow time-varying and space-varying coherent multipath channel". Aiming at the problem of active sonar target recognition, Chap. 7 further studies the characteristics of active sonar target channel. In order to fulfil the function of "Bridge the gap between acoustic and underwater engineering", the last chapter gives a typical example of applications close integration with today's sonar. It provides samples so as to lead the readers to the concept of acoustic channel.

This book is a monograph about the underwater acoustic channel. It can provide study materials and references for the technical staff in the field of underwater acoustic positioning navigation, the sensor array design, underwater target features, underwater acoustic image and communication, noise measurement and control, and the vector acoustic. Moreover, it can be used as textbooks and reference books for senior and postgraduates majoring in underwater acoustics studying in universities and scientific research institutes. The book strives to be concise, keeps the mathematics simple, and collects substantial experimental data. Thus, the readers can easily understand the basic ideas. It is hoped this book will benefit readers in their study and research.

This book is co-edited by Profs. Hui Junying and Sheng Xueli of Harbin Engineering University. Professor Hui Junying is the editor-in-chief of this book and edits the final compilation. Chapters on sound pressure time-reversal mirror and vector time-reversal mirror are edited by Prof. Sheng Xueli.

Some of the first drafts of this book were supervised by Academician Yang Shie of Harbin Engineering University and the late Prof. Zheng Zhaoning of the Naval Academy of Engineering. Professor Yin Jingwei, Prof. Mei Jidan, Dr. Yang Juan, and Dr. Yao Zhixiang have worked hard for the review of the second edition of this book.

Many thanks go to Prof. Zhou Weiwei from School of Foreign Studies, Harbin Engineering University and Dr. Cang Siyuan from College of Underwater Acoustic Engineering, Harbin Engineering University, for their contribution to the translation work of this book.

The authors would like to express their sincere thanks to the special fund in the "Tenth Five-Year Plan" on the construction of postgraduate teaching materials of Harbin Engineering University, for the funding on the compilation and publication of this book.

Harbin, China   Junying Hui
                                                                                   Xueli Sheng

Content Brief Introduction

This book builds a "bridge" between underwater acoustic physics and sonar engineering design. In terms of communication theory, it discusses the description method, characteristics of the underwater acoustic channel, and its influence on underwater acoustic systems in an in-depth and systematic way with the mathematical method that engineers and technicians can understand. The book consists of nine chapters. The main contents are basic review and sonar equation, the basic concept of the ocean acoustic and vector acoustic, the average energy channel, coherence multipath channel, random time-varying space-varying channel theory foundation, slow time-varying coherence multipath channel characteristic, reverberation channel, etc. For the active sonar target recognition problem, this book also further studies the active sonar target channel characteristic. In order to give readers a deeper understanding of the knowledge of the book, a typical example closely integrated with today's sonar applications is presented at the end of the book. This book collects extensive data from sea trials to illustrate physical images of underwater acoustic channel transmission characteristics.

This book is a monograph about the *Underwater Acoustic Channel*. It can provide study materials and references for the technical staff in the field of underwater acoustic positioning navigation, the sensor array design, underwater target features, underwater acoustic image and communication, noise measurement and control, and the vector acoustic. Moreover, it can be used as textbooks and reference books for senior and postgraduates majoring in underwater acoustics studying in universities and scientific research institutes.

Contents

Chapter 1
Introduction

1.1 Introduction

The ocean is regarded as a blue gem on earth. It is predicted that people will rely more on marine resources, obtain more food, energy, and minerals, as well as exploring the earth's secrets from the ocean in the twenty-first century.

Since people realized that sound wave is the farthest physical field in the ocean, it has become the main tool to study and explore the ocean. Now, with the development of marine exploration and military needs, underwater acoustic technology has risen to be an important industry and a new star in the field of high technology. Underwater acoustic technology has been widely used in navigation, underwater observation, underwater communication, fishery, ocean development, seabed resources investigation, and marine physics research. Especially in military application, it serves as the key technology in submarine operation and anti-submarine operation, mine operation, and anti-mine operation. At present, many countries in the world are competing to develop underwater acoustic technology.

Being different from acoustics with a long history, hydroacoustics is a newly-developed modern science. Two centuries have passed since D. Colladon and C. Sturm skillfully measured the velocity of the sound wave in water for the first time in Genfer lake in the early eighteenth century. At first, attention was directed to underwater acoustic technology only for military needs. In the first World War (1914–1918), the Triple Ententelost one-third of the total tonnage of ships because of German submarines. Since then, people began their study of how to employ sound waves to detect submarines. From 1916 to 1918, Langevin, a famous French physicist, and chlosky, a Russian Engineer, developed an active sonar device, which successfully received the echo from a submarine under the water of 1500 m away. However, sonar was not used in the first World War. After that, sonar technology entered the application stage, thanks to the development of electronic technology and electroacoustic transducer. During the Second World War, the underwater acoustic equipment has been improved, which plays an important role in the submarine operation and

© Harbin Engineering University Press 2022
J. Hui and X. Sheng, *Underwater Acoustic Channel*,
https://doi.org/10.1007/978-981-19-0774-6_1

anti-submarine operation. In fact, underwater acoustic technology was rapidly developed after World War II. Low frequency, high power, and large array became the trend of sonar technology development at that time, especially the mature research on the surface channel, submarine reflection sound, deep-sea channel, and acoustic convergence area effect, making the sonar's working distance increase in the early 1960s. In the early 1950s, the matching filtering technology was successfully applied in radar technology, resulting in the leap of its working distance, and stimulating people to track matching filtering technology in sonar technology in the next 20 years. In fact, no good effect was reflected in the radar technology field. People realized that underwater acoustic channel is far more complex than radar channel, thus the theory of acoustic channel had been concerned in the early 1970s. Low frequency, large array and sonar system which takes high-speed computer as the center to obtain the information of ocean acoustic channel and self-adaptive processing in real-time have become a new trend of sonar technology development.

Sonar transmitting transducer array or sound source sends out the sound wave with information. The sound wave propagates in the ocean. Then, it is received by the hydrophone array. The Sonar system processes the received signal and then determines whether there is a target and what its state parameters are, as well as the types of target, or recover the source information sent by the target. This is the whole process of the sonar system. From the perspective of communication theory, the ocean is the sound channel. The ideal channel can transmit information without distortion. The ocean proves not to be ideal, but complex and changeable. Only by fully understanding the limitation of ocean acoustic channels on the sonar system, can people be free to gradually adapt the sonar system to the ocean environment, in order to obtain better detection effect and recognition ability. Ocean acoustic channel not only makes energetic transformation on target radiation signal but also performs information transformation. The coherence multi-path structure of the underwater channel will distort the received signal waveform, which is significantly different from the source radiation waveform. Ocean acoustic channel is more complex because of its random variation with time and space. In the process of transmission, information is transformed and lost at the same time. The theory of acoustic signal will study the transformation of information by the channel and how the sonar system adapts to the acoustic channel.

The authors have noticed the following problems: Although most underwater acoustic engineers have sufficient knowledge of signal processing and system design, they do not fully understand the influence of complex marine environment on sonar design; Similarly, although marine physicists and underwater acoustic physics workers have extensive knowledge of the marine environment and its relevant research, they have little knowledge about the relationship between these knowledge and sonar system work. Therefore, this book attempts to bridge a communication channel between the two sides. The book is dedicated to sonar engineers, senior undergraduates, and postgraduates majoring in underwater acoustics. It introduces the limitation of the marine environment on sonar system performance and discusses

the adaptation of the two with the language and mathematics easily accepted by engineers and technicians. This book is also of reference value to technicians engaged in
research in the field of radar and communication.

1.2 Mechanical Vibration

It has always been saying, "vibration makes the sound", which means that sound wave
stems from vibration, and sound wave results from the propagation of vibration in
the medium.

The so-called vibration refers to the reciprocating motion of the intermediate point
around the equilibrium point.

Since a practical vibration system is highly complicated, how shall we make
studies on it? Physicists state that no science is without models, and the study of any
actual physical problem should be abstracted into a physical model. Any valuable
physical model features two points: one is to be as simple as possible so that it can be
analyzed with simple mathematical tools; the other is that it must be a duplicate of
the real physical problem, that is, it must include the main contradiction of the actual
problem. Only under certain experimental conditions can the model be verified to
be correct and valuable. Any vibration system in a narrow frequency band can be
abstracted as a simple vibration model of "single degree of freedom particle vibration
system". The purpose of studying this simple vibration model in this section is to
clarify the basic concepts of vibration.

Figure 1.1a is a schematic diagram of a single degree of the freedom vibration
system. A steel ball m is hung under the spring D. The spring is an elastic element and
the steel ball is a mass element. The balance position of the steel ball is taken as the
origin of the coordinate axis X. If for some reason, the ball has an initial displacement,
the vibration will be motivated, that is, the ball will move back and forth around
the equilibrium position. Figure 1.1b is the structural diagram of an underwater
acoustic transducer, which can be abstracted as a single degree of freedom vibration

Fig. 1.1 Single degree of
the freedom vibration system

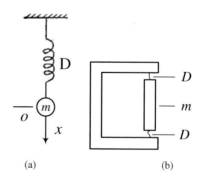

(a)   (b)

system model. The characters in the figure are equal to the equivalent elements of corresponding actions.

When the steel ball deviates from the equilibrium position and has a displacement x, the spring is compressed or stretched. The elastic force produced by the spring, which also acts on the steel ball, is f, whose size is directly proportional to the displacement, and its direction is opposite to the displacement direction, namely,

$$f = -Dx \qquad (1.1)$$

where the proportional constant D is the elastic coefficient and its reciprocal is the compliance coefficient:

$$C_M = 1/D$$

Ignoring the influence of mass and gravity of spring, and based on Newton's second law, the equation of motion of steel ball can be obtained as follows:

$$\frac{d^2x}{dt^2} + \omega_0^2 x = 0 \qquad (1.2)$$

$$\omega_0^2 = D/m$$

where m means the mass of the steel ball; ω_0 represents angular resonance frequency of vibration system, and the solution of Eq. (1.2) is as follows:

$$x(t) = C_1 \cos \omega_0 t + C_2 \sin \omega_0 t$$
$$= C \cos(\omega_0 t + \varphi) \qquad (1.3)$$

$$C = \sqrt{C_1^2 + C_2^2}, \quad \varphi = -arctg(C_2/C_1)$$

Accordingly, when the steel ball vibrates harmoniously, the steel ball moves back and forth around the equilibrium position following the law of sine or cosine. C is called amplitude and φ is called the initial phase.

It is of great significance to introduce the concept of electromechanical analogy, especially for readers who are familiar with the theory of electrical oscillation. Single degree of freedom vibration system and single resonant circuit have the same form of the differential equation, though their physical properties are diverse, the characteristic quantities of mechanical vibration and electrical oscillation have the same function formulation. They follow the same mathematical relationship, so there is a possibility of making comparisons. Some relations of electro-mechanical analogy are illustrated as follows, shown in Fig. 1.2.

$x(t)$ is displacement.

$v(t) = \frac{dx}{dt}$ (vibration velocity) $\leftrightarrow i(t)$ (electric current)

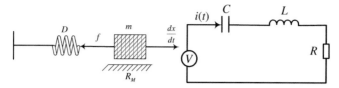

(a) Single degree of freedom vibration system   (b) Single oscillation circuit

Fig. 1.2 The diagram of electromechanical analogy

$f(t)$(force) $\leftrightarrow$ $V(t)$(voltage)

m(quality) $\leftrightarrow$ L(inductance)

C_M(compliance coefficient) $\leftrightarrow$ C(capacitance)

$f(t) = Fe^{j\omega t} \leftrightarrow V(t) = Ve^{j\omega t}$

$\frac{dx}{dt} = V_M e^{j\omega t} \leftrightarrow i(t) = Ie^{j\omega t}$

$Z_M = F/V_M$(mechanical resistance) $\leftrightarrow$ $Z = V/I$(electric resistance)

$Z_M = R_M + j\left(m\omega - \frac{1}{\omega C_M}\right) \leftrightarrow Z = R + j\left(\omega L - \frac{1}{\omega C}\right)$

$P_M \quad = \quad \frac{1}{2}|F||V_M|\cos\varphi_M$(mechanical power)$\leftrightarrow P \quad =$

$\frac{1}{2}|I||V|\cos\varphi$(electric power)

ϕ_M—phase difference between force and vibration velocity$\leftrightarrow$

ϕ—phase difference between voltage and current

$Q_M = \frac{\omega_0 m}{R_M}$ (mechanical quality factor) $\leftrightarrow$ $Q = \frac{\omega_0 L}{R}$ (electrical quality factor.)

1.3 Basic Concepts of Sound Wave

This section briefly discusses the basic concept of the sound wave, which is dedicated to non-sound professional readers or learners. The dictions are not elegant but popular.

The vibration's propagation in mediums is called sound wave. The vibration source is the sound source. The simplest sound source is a uniform pulsating sphere, on which each point vibrates harmoniously with the same vibration velocity and the same phase. The direction of vibration velocity points to the radiation direction, that is, the direction of vibration velocity is perpendicular to the spherical surface. When the medium is disturbed by the vibration of the sound source, the points in the medium appear to vibrate harmoniously, and the medium at each point is compressed or stretched (sparse). The overpressure produced by the pressure of the medium is called sound pressure. The propagation speed of the vibration state in the medium is called sound speed. For harmonic sound waves, phase can be used to represent the state of vibration, i.e., if the velocity of vibration on an infinitesimal uniform pulsating sphere is assumed to be

$$v(t) = V_M e^{j\omega t}$$

Considering that the sound source and medium are spherically symmetric, it is easy to understand that sound waves should also be spherically symmetric. Since the medium mass at the distance r repeats the vibration state of the sound source at time t with the lag time r/c, and c is the propagation velocity of the sound wave in the medium, the vibration velocity at the distance r can be written as:

$$v(r, t) = V_M(r)e^{j\omega[t-(r/c)]} \tag{1.4}$$

where $V_M(r)$ shows the amplitude of sound wave changes with distance r, and the factor $\omega\left(t - \frac{r}{c}\right)$ represents the phase of medium particle vibration. The equal phase plane is called wave front. The wavefront represented in Eq. (1.4) is spherical. For a given r, each point on the sphere has the same vibration phase, that is, it has the same vibration state. The propagation velocity of wave front is identical to the phase velocity of a sound wave, which is simply put as sound velocity.

As we all know, there are two kinds of theories about light propagation: ray theory and wave theory. According to the ray theory of light, the energy of light propagates along the ray, and the ray is a straight line in a uniform medium. The ray theory of sound propagation is briefly described below. According to the sound line theory, the sound energy propagates along the sound line, which is perpendicular to the wave front. A series of sound lines form the sound beam tube. The sound energy emitted from the sound source propagates along the sound beam tube in the lossless medium, and its total energy remains unchanged, so the sound intensity is inversely proportional to the cross-sectional area of the sound beam tube.

Now we use the ray theory to study the sound field of a pulsating sphere. It has been shown that the wave front of the pulsating sphere is a series of concentric spheres. The sound line is a series of radiation lines emitted by the sound source, which are perpendicular to the wave front, as shown in Fig. 1.3. Therefore, with the increase of the distance r, the cross-sectional area of the acoustic tube increases according to its square law, and the sound intensity decreases according to its square law. The reduction of sound intensity due to the expansion of wave front is called "geometric loss", which is called spherical wave attenuation law. Since the sound intensity is

Fig. 1.3 Sound field line and wave front of pulsating sphere

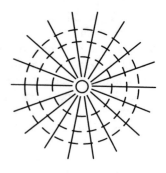

inversely proportional to the square of the distance, the vibration velocity is inversely proportional to the distance.

$$v(r, t) = \frac{A}{r} e^{j\omega\left(t - \frac{r}{c}\right)} \tag{1.5}$$

where A is a constant variable, which depends on the power of the sound source.

The physical quantities describing the sound vibration at a certain point in the harmonic sound field include sound pressure, vibration velocity and sound power. The parameters describing the sound vibration include frequency, amplitude and phase.

The wave array surface is called plane wave, and the wave wave with the surface of the wave array is called spherical wave. The following is a list of their basic physical quantities.

Physical variables	Spherical wave	Plane wave
Vibration velocity	$v(r, t) = \frac{A}{r} e^{j(\omega t - kr)} =$ $V(r) e^{j(\omega t - kr)}$	$v(x, t) = A e^{j(\omega t - kx)} = V e^{j(\omega t - kx)}$
Sound pressure	$p(r, t) = \frac{B}{r} e^{j(\omega t - kr)} =$ $P(r) e^{j(\omega t - kr)}$	$p(x, t) = B e^{j(\omega t - kx)} = P e^{j(\omega t - kx)}$
Sound intensity	$I(r) = \frac{1}{2}|V(r)||P(r)| =$ $\frac{|P(r)|^2}{2\rho c}$ $= \frac{1}{2}\rho c |V(r)|^2 \propto \frac{1}{r^2}$	$I = \frac{1}{2}|V||P| = \frac{|P|^2}{2\rho c}$ $= \frac{1}{2}\rho c |V|^2$

In this table, A and B are constants, depending on the power of the sound source. ρ and c are medium density and sound velocity respectively.

$$k = \omega/c = 2\pi/\lambda, \lambda = c/f$$

where k is called the wave number, λ is the wavelength, and f is the frequency of the sound wave. The units of sound pressure are listed below:

$$1 \, (\text{Pa}) = 1 \, (\text{N}/\text{m}^2)$$

$$1 \, (\mu b) = 10^{-5} (\text{N}/\text{cm}^2)$$

$$1 \, \text{Pa} = 10 \, \mu b$$

$$1 \, \mu b = 10^5 \, \mu \text{Pa}$$

In the air, the human hearing threshold (audible level) for 1000 Hz pure tone is about 2×10^{-5} Pa or $2 \times 10^{-4} \mu b$.

Usually, the sound pressure is about 0.1 Pa when talking loudly indoors. In water, the sound pressure of weak signal received by underwater acoustic equipment is about $1 \times 10^5 \mu$Pa. When a pound of TNT explodes underwater, the peak sound pressure at 100 m is about 2×10^5 Pa. In the air, the vibration velocity corresponding to 1 Pa sound pressure is about 2.4×10^{-3} m/s. The corresponding vibration velocity in water is about 7×10^{-7} m/s. In terms of air, $\rho = 1.29$ kg/m^3, $c = 331$ m/s, and the acoustic resistance appears to be $\rho c = 430$ kg/s m^2; For water, $\rho = 1 \times 10^3$ kg/m^3, $c = 1500$ m/s, and the acoustic resistance becomes $\rho c = 1.5 \times 10^6$ kg/s m^2.

In acoustics, sound level is often used to indicate sound intensity or pressure. Its is defined as:

$$IL = 10 \log I / I_0 = 20 \log P / P_0 (\text{dB})$$

where P_0, I_0 refers to referential sound pressure and intensity. P_0 is the reference level in aeroacoustics. Its value is $P_0 = 2 \times 10^{-5}$ Pa. In underwater acoustics, according to the current international standards, $P_0 = 1 \mu$Pa. Accordingly, the sound pressure 0.1 Pa in water can be defined with the sound intensity level of 100 dB. Historically, $P_0 = 1 \mu$b has ever been used in many kinds of literature, which deserves readers' special attention. This book is subject to the current international standards. Otherwise, it will be specially explained.

1.4 Ohm's Law of Acoustics

It is well known that the ratio of voltage added to a resistor to the current passing through the voltage is constant, and its ratio coefficient is the resistance value, such as

$$\frac{V(t)}{i(t)} = R \tag{1.6}$$

It is called Ohm's law. If the inductance and capacitance are found in circuit, Eq. (1.6) can be expressed as:

$$\frac{V(t)}{i(t)} = Z(\omega) \tag{1.7}$$

where $Z(\omega)$ is named as complex impedance. For the single oscillation circuit shown in Fig. 1.2b, we obtain

$$Z(\omega) = R + jX = R + j\left(\omega L - \frac{1}{\omega c}\right) \tag{1.8}$$

Equation (1.6) shows if the impedance is real number, the waveform of voltage is obtained as the current waveform is multiplied by a constant coefficient, both of which have the same waveform; For sinusoidal oscillation, voltage and current are in the same phase. If the impedance is complex number, the voltage and current have phase difference, or for broadband signal, the waveform of both will be different.

The following is about some issues in acoustics.

The harmonic sound wave, that is, the time function of the sound wave with $e^{j\omega t}$, is studied. The acceleration $\boldsymbol{a}$ serves as the time derivative of the vibration velocity $\boldsymbol{v}$, such as

$$\boldsymbol{a} = j\omega\boldsymbol{v} \tag{1.9}$$

According to Newton's Law, we obtain

$$-\nabla p = j\omega\rho\boldsymbol{v}$$

$$\nabla p = -j\omega\rho\boldsymbol{v} \tag{1.10}$$

The sound pressure gradient $<-\nabla p>$ on the left side of the above formula represents force, ρ is the density of the medium and ω shows the angular frequency.

If the sound wave is a harmonic plane wave, the plane wave pressure $p(x, t)$ listed in the table of Sect. 1.3 can be:

$$p(x, t) = Pe^{j(\omega t - kx)} \tag{1.11}$$

where p is the amplitude of sound pressure, the wave number appears to be $k = \omega/c$, and c means the sound velocity of the medium.

When putting the above formula into (1.10), we can obtain

$$\frac{p(t)}{v(t)} = \rho c \tag{1.12}$$

which is called Ohm's Law of plane sound waves. The waveforms of both sound pressure and vibration velocity, emerging in the plane sound waves, are identical and completely related.

For spherical sound waves,

$$p(r, t) = \frac{P}{r}e^{j(\omega t - kr)} \tag{1.13}$$

Fig. 1.4 The phase
difference between sound
pressure and velocity in a
spherical wave field

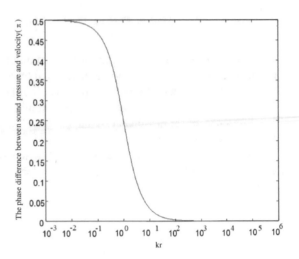

It is obtained from the table in the previous section, where P represents the sound
pressure amplitude at 1 m.

Substituting Eq. (1.13) into Eq. (1.10), the above formula takes r for derivative
and obtains

$$\frac{p(r,t)}{v(r,t)} = Z(\omega) = \frac{\rho c}{1 - j\frac{\lambda}{2\pi r}} \qquad (1.14)$$

It can be seen from the above that the impedance of spherical wave is complex at
a short distance, and there is a phase difference between sound pressure and vibration velocity. For broadband signal, the waveforms of sound pressure and vibration
velocity will result in different forms. But in fact, as long as $r > \lambda$, the imaginary
part of Eq. (1.14) can be negligible (see Fig. 1.4), its impedance is approximately
equal to the acoustic impedance ρc of plane wave, and the normalized correlation
coefficient of sound pressure and vibration velocity is almost to be 1.

Any complex traveling wave sound field can be approximated as a plane wave
in the distance, so their acoustic impedance is a real number. Although the sound
field of the point source in the ocean is very complicated, the sound pressure and the
vibration velocity are almost in phase at several times the distance of the sea depth,
and both are completely related.

1.5   Basic Concepts of Vector Acoustics

Both sound pressure and vibration velocity exist in the sound filed, in which the
former is scalar field and the latter is vector field. For more than one hundred years,

hydrophone is used to pick up the sound field information in water. As a sound pressure sensor, hydrophone is different from seismic acoustics. Underwater acoustics keeps on paying close attention to the scalar field sound pressure, but has neglected the vibration velocity field for nearly a century.

Rayleigh disk was the first device to measure the vibration velocity in fluid. In 1942, Bell Telephone laboratory developed the first moving coil vibration velocity sensor. In 1958, G. L. Boyer successfully developed the sound pressure gradient velocity sensor. It was only 20 years ago that the vibration velocity sensor technology made a rapid advancement, and the shelf products applied in engineering appeared. Vector acoustics and vector acoustic signal processing technology sprang up gradually.

The acoustic involving the research of sound pressure and vibration velocity is called "vector acoustics".

A combined sensor of sound pressure and vibration velocity outputs sound pressure p and three-dimensional vibration V_x, V_y, and V_z at the same point, listed as

$$\begin{cases} V_x = V \cos \theta \cos \alpha \\ V_y = V \sin \theta \cos \alpha \\ V_z = V \sin \alpha \end{cases} \tag{1.15}$$

The mode of vibration velocity in the above formula is V, the horizontal azimuth θ is $0°$ on the X axis, and elevation α is $0°$in horizontalplane。 For plane sound, if the sound pressure is:

$$p(t) = x(t)$$

According to Ohm's Law of acoustics, we obtain

$$\begin{cases} V_x(t) = \frac{1}{\rho c} x(t) \cos \theta \cos \alpha \\ V_y(t) = \frac{1}{\rho c} x(t) \sin \theta \cos \alpha \\ V_z(t) = \frac{1}{\rho c} x(t) \sin \alpha \end{cases} \tag{1.16}$$

Considering signal processing, in order to write easily and not to lose generality, the acoustic impedance on the right side of equation is omitted, and the constant coefficient on the right side of formula (1.16) is 1. Sounder the plane wave conditions, there are:

$$p(t) = x(t) \quad V_x(t) = x(t) \cos \theta \cos \alpha$$
$$V_y(t) = x(t) \sin \theta \cos \alpha \quad V_z(t) = x(t) \sin \alpha \tag{1.17}$$

Since the sound pressure is scalar, a small-scale sound pressure sensor can be non-directional; The vibration velocity is a vector, so a vibration velocity sensor has dipole directivity, as shown in the above formula and Fig. 1.5. The directivity of vibration

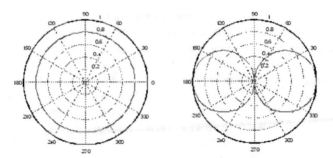

a. directivity of sound pressure sensor   b. directivity of vibration velocity (VX) sensor

Fig. 1.5 Directivity of vector sensor

velocity is independent of frequency, so even for a few hertz VLF sound wave, a small vector sensor has directivity. In low-frequency band, one of the advantages of vector signal processing is to use vector sensor to determine the direction of sound source.

The directivity can be rotated in three-dimensional space by using the weighted linear combination of three velocity components. Take horizontal rotation directivity as an example:

$$V_c(t) = V_x(t) \cos \psi + V_y(t) \sin \psi$$
$$V_s(t) = -V_x(t) \sin \psi + V_y(t) \cos \psi \quad (1.18)$$

In the above formula, ψ is called the guiding direction, which is the principal maximum direction of V_c and the directivity zero-point direction of V_s. Taking $\alpha = 0$ from Eq. (1.17) and substitute it into Eq. (1.18), we obtain

$$V_c(t) = V(t) \cos(\theta - \psi)$$
$$V_s(t) = V(t) \sin(\theta - \psi) \quad (1.19)$$

Under the condition of plane wave, the above formula turns to be:

$$V_c(t) = x(t) \cos(\theta - \psi)$$
$$V_s(t) = x(t) \sin(\theta - \psi)$$

As long as the value ψ is changed, the dipole directivity can rotate in space. It is not difficult for readers to ponderover the method of directivity rotating in the vertical plane.

The sound intensity $|I|$ of plane wave is given in the table in Sect. 1.3 as follows:

$$|I| = \frac{1}{2}|V||p| = \frac{|p|^2}{2\rho c} = \frac{1}{2}\rho c |V|^2 \quad (1.20)$$

Fig. 1.6 The average sound
intensity device

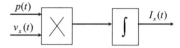

The instantaneous sound intensity current I is as follows:

$$I = p(t)V(t) \tag{1.21}$$

The sound intensity current I is a vector whose direction matches with the direction of sound energy. The time average value of its mode is the sound intensity, and its average value is the average sound intensity current.

Figure 1.6 is the principleframe of the average sound intensity device. It consists of a multiplier and a time integrator.

In the isotropic acoustic interference field, the interference noise arrives from all directions, with the same intensity in the sense of statistical average. The average sound intensity current of the interference at the receiving point is the synthesis of the interference sound intensity current vectors arriving from all directions, and the expected value of the synthesis vector must be zero, i.e., in the isotropic interference noise field, the output interference background of the average sound intensity detector is very small, but with liable anti-interference ability.

If the target signal sound field is plane wave and the interference is isotropic, the average signal intensity of the target signal output by the sound intensifier is the target sound intensity, and the interference background is very weak.

1.6   Sonar System and Acoustic Channel Model

Sonar system and ocean acoustic channel are so complicated that they must be abstracted into a model for further study. The proposed model, clarifying the main contradiction in sonar work, embodies people's understanding of sonar system and ocean channel.

Sonar transmitting transducer array or sound source sends out sound wave carrying information, and arrives at the receiving hydrophone array through the ocean. Sonar system analyzes and processes the received signal for determining whether there is a target, and what the parameters and types of the target are, which constitutes the whole working process of sonar. In view of communication theory, the ocean is regarded as the sound channel. The ideal channel can transmit information without distortion. The ocean is the channel which is not ideal, but complex and changeable. Only by fully identifying the influence and limitation of acoustic channel on sonar system design, can people be free to gradually adapt sonar system to acoustic channel, so as to obtain good detection effect and recognition ability.

The information flow of active sonar is similar to that of radar, as shown in Fig. 1.7. The signal source generates signal $z(t)$, which is converted into an acoustic signal by

Fig. 1.7 Schematic diagram
of active sonar system and
acoustic channel

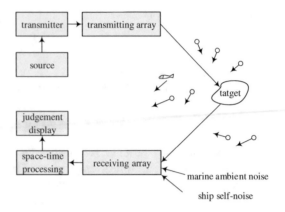

the transmitting array. The acoustic wave reaches the target through the ocean, and
the scattered acoustic wave (echo) generated by the target arrives at the receiving
hydrophone array through the ocean again. The blue ocean has heterogeneous characteristics in acoustic. There are bubbles, suspended particles, plankton, fish and
shrimp, temperature microstructure, water masses, turbulence and internal waves in
the ocean; The bottom of the sea and the sea are not even. The phenomenon that
the sound energy is re-radiated when it encounters these inhomogeneous bodies is
called scattering. Reverberation is an important interference of active sonar. In the
process of sound wave propagation in the ocean, not only its sound intensity gradually decreases, but also the acoustic signal waveform distortion is caused by the
above-mentioned non-uniformity and acoustic interference phenomenon, especially
the multiple reflection of sound waves from randomly moving uneven boundaries,
leading to the acoustic signal random distortion and information loss. These distor-
tions limit the detection performance of sonar system. The relative motion between
the target and the receiving and transmitting arrays and the complexity of the scattering characteristics of the target mainly result in the distortion of the acoustic signal.
Many kinds of sound sources abound in the ocean, with which the so-called marine
environmental noise is formed. The carrier of sonar also has strong self-noise which
usually covers up the echo together with reverberation from a long distance. The
spatio-temporal processor of sonar system analyzes the received echo and interference, followed by the decision of whether there is a target and estimates the state
parameters and properties of the target.

From the point of view of communication theory, the ocean can be regarded as a
random time-varying and space-varying filter, which transforms the signal generated
from the sound source. Except for the nonlinear phenomenon of acoustics, the ocean
medium (including boundary and target) between sound source and receiver can be
reasonably assumed as a linear filter.

In summary, the sonar system and acoustic channel can be represented by the
model shown in Fig. 1.8. The target is regarded as a part of the channel, which
transforms the incident acoustic signal so that the echo fetches the information of
the target.

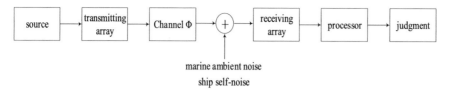

Fig. 1.8 The model of the sonar system and acoustic channel

The information flow and its model of passive sonar is similar to that of active sonar. In this case, the sound source is the target of information source, and the marine environment noise and ship self-noise are considered as the main interference.

This book is devoted to discussing the characteristics of ocean channel, which is regarded as a time-varying and space-varying filter, studies the description method of its characteristics, describes their system functions and their physical meaning, and explores the method of matching sonar system with acoustic channel.

Acoustic channel theory is the weakest research point in sonar technology. Early studies on acoustic channel focused on the propagation mode and average energy loss. Generally speaking, it is to understand under what conditions the sound can spread further in the ocean, so as to enable the sonar access to a greater range. At present, the research in this area has been relatively mature. With the discovery of the working mode of convergence area and submarine reflection, as well as the understanding of sound absorption, the operating range of sonar has been greatly improved in 20 years after the Second World War. Due to the profound study of ship noise line spectrum and the development of signal processing technology, towed linear array sonar emerges accordingly. Also, because it adopts line spectrum detection technology and large array, its operating range has achieved a second leap.

The classical detection theory is based on the following model: In the band of interest, the interference appears to be stable limited white Gaussian noise, the acoustic channel is an ideal channel, and the target is a point target. The main result of classical detection theory is matched filter theory, and its most famous practice in sonar technology is copy correlator. In the past 20 years, many kinds of copy correlators designed under the guidance of the classical theory have proved that its practice effect is not ideal, mainly because the sonar channel is beyond satisfactory. Firstly, the reverberation is not white interference or even stable; Secondly, the multipath channel distorts the acoustic signal, so the detected signal waveform is unknown. The delicately designed waveform with ideal ambiguity function (thumbtack function) dramatically increases the sidelobe of the received signal ambiguity function due to the multipath effect. In a word, the classical matched filter theory has proved to be not sufficient, and a new detection method needs exploring, especially the characteristics of the acoustic channel.

The purpose of this book does not mainly focus on making a comprehensive and in-depth introduction to the development of modern underwater acoustics. Although we need to explore new sonar technology theory, its systematicness is not the study target. The book aims to make a preliminary description of random acoustic channel and points out its influence on sonar design. We hope to bridge the gap between underwater acoustic physics, underwater acoustic signal processing and underwater acoustic engineering. The basic theory of channel introduced in this paper provides the opportunities for readers to read relevant literature further.

1.7   Sonar Equation

Similar to other detection problems, the basic contradiction in sonar work is that the signal must be detected in the interference background, and the interference limits the decision of the signal. There is no distinction without differences. According to the amplitude difference of interference and signal, or the difference of their statistical characteristics in time and space, it can be seen whether there is a target or only interference exists. For passive sonar, signal is the noise coming from the target; For active sonar, signal is the echo reflected by the target. Ship's self-noise and marine environment noise are regarded as background interference, and a special interference reverberation features for active sonar. The basic knowledge of active sonar design is to make the radiated sound power large enough so that the ocean reverberation becomes the main interference limiting the detection under the condition of Engineering permission. Therefore, reverberation is usually one of the important interferences of active sonar. Sonar equation synthesizes both the combination of target signal and interference contradiction, which is used to estimate the influence of various design parameters and environmental factors in sonar system model on sonar performance, mainly on operating range. Taking active sonar as an example, the factors in sonar system model are as follows.

Sound source: the sound source level SL is determined by the directivity concentration coefficient and sound power of the sound array. The waveform $s(t)$ and the spectrum of radiation signal $S(f)$ are determined by signal generator and acoustic array.

The acoustic channel transforms the average energy and information of the acoustic signal: the acoustic channel is regarded as a time-varying and space-varying random filter, but it can be seen as a slow time-varying filter in most applications (this will be explained in the following chapters).

Spatiotemporal joint processing: the structure of transmitting array, receiving array and sonar signal processor as well as the spatiotemporal statistical characteristics of acoustic channel will determine the detection threshold of sonar system and the detection and decision ability of sonar system.

The classical sonar equation is based on the description of the average energy of the sonar system model. Though it is elementary, and the sonar parameters in the equation are scattered in a large range, its application has obvious limitations, but the classical sonar equation can simply explain the interrelationship between the factors that affect the sonar operating distance, which is still a start to discuss sonar technology.

The active sonar equation is as follows:

$$SL - 2TL + TS - (NL - DI) - RL \geq DT \qquad (1.22)$$

The following is for passive sonar:

$$SL - TL - (NL - DI) \geq DT \qquad (1.23)$$

The meaning of each variable in the formula is as follows:

SL
: sound source level: the sound intensity level at 1 m away from the acoustic center of the radiation source.

TL
: one way propagation loss of sound wave from sound source to target (or observation point).

TS
: backscattering intensity of target: the decibels of the ratio of reflected sound intensity to projected sound intensity at 1 m from the acoustic center of the target.

Sl-2tl + TS
: echo signal intensity level at the receiving hydrophone.

NL
: the ship self-noise and environmental noise level received by the receiving hydrophone array.

DI
: directivity index of receiving hydrophone array.

NL-DI
: Ship self-noise and ambient noise level received by directional hydrophone array.

RL
: reverberation level.

DT
: detection threshold.

Detection threshold refers to the given decibels of the lowest signal to interference power ratio at the receiver hydrophone for a given false alarm probability P_α and detection probability P_d when the false alarm rate of the sonar system is not more than P_α and the detection probability is not less than P_d. The detection threshold is determined by the performance and signal of array and processor, as well as the spatiotemporal statistical characteristics of signal and interference. On the left side of the sonar equation is the output power signal to clutter ratio (DB) of the array which varies with the target distance. For a certain limit operating distance, the sonar equation becomes an equation. In other words, the sonar equation indicates that only when the power signal to clutter ratio of the sonar receiving array is greater than or equal to the detection threshold, can the sonar work normally with a confidence limit (P_α and P_d) better than the predetermined limit.

Sonar equation is elementary, but very useful in engineering design. This book will explain the physical meaning of parameters in sonar equation and the basic concepts of engineering application.

The purpose of the application of sonar equation is to make it easy for readers to understand the meaning of sonar equation.

If the reverberation interference can be ignored and the range of sonar is limited by local noise and ambient noise, then Eq. (1.22) can be rewritten as follows:

$$SL - (NL - DI + DT) + TS = 2TL$$

$$GL = SL - (NL - DI + DT) + TS \qquad (1.24)$$

The passive sonar equation can also be rewritten as:

$$SL - (NL - DI + DT) = TL$$
$$GL = SL - (NL - DI + DT) \qquad (1.25)$$

GL is referred to as the high-quality factor of sonar system. For passive sonar, the meaning of sonar equation is that the maximum one-way propagation loss allowed shall not exceed the system quality factor; For active sonar, the maximum allowable two-way propagation loss shall not exceed the high-quality factor of sonar system after deducting the target strength. $TS = 0$ is taken when calculating the high-quality factor. In other words, the larger the quality factor of the system, the farther the distance of the system. The higher quality factor is one of the characteristics used to evaluate the merits and disadvantages of sonar system. The high-quality factor can be determined by measuring various design parameters of sonar system in laboratory and offshore test, or by measuring propagation loss simultaneously in the system offshore test. The high-quality factor is an index to evaluate the system and is also an indispensable quantity for predicting the sonar range. In the marine identification test of sonar system, the high-quality factor of the appraisal system should be more universal than that of the evaluation under specific hydrological conditions and target performance.

A simple analysis of Eq. (1.24) and Eq. (1.25) can point out a basic direction of sonar development, that is, the more the electro acoustic performance of sonar and the acoustic performance of the ship are improved, i.e., the higher the quality factor of sonar is, the lower the working frequency of sonar should be.

Taking the passive sonar as an example, assuming that the speed of the ship is very low and the ship is low-noise, the main interference is the marine environment interference, which can be represented in Eq. (1.25) as follows:

$$TL = SL - (NL - DI + DT)$$

where the propagation loss TL includes the geometric attenuation and physical absorption caused by the expansion of the wave front. The sound energy propagates along the acoustic tube, and certain amount of sound energy is distributed on the larger section of the acoustic tube. The sound energy per unit area becomes smaller and smaller, and the sound intensity decreases with the expansion of the wave front. In the process of sound propagation, sound energy is gradually transformed into heat energy, which is called physical absorption, and the corresponding reduction of sound intensity is called absorption loss. The loss of communication can be written as follows:

$$TL = 15 \log r + 0.036 f^{3/2} \cdot r + 60 \text{ (dB)}$$

where the unit of r is km and the unit of frequency is kHz. The middle term on the right represents the physical absorption loss, which is based on the empirical formula obtained from experiments; The other two items are geometric loss, in which r should be greater than 1 km when using this formula. It is assumed that the attenuation of spherical wave is within 1 km, that is, the geometric loss is 60 dB at 1 km. When the distance is greater than 1 km, the geometric loss is between the spherical law (TL = 20logr) and the cylindrical law, and a 3/2 power law can be obtained.

The interference level received by the receiving array is as follows:

$$NL - DI = NL_0 - 10c \log f - 10 \log \gamma_0 - 20 \log f + 10 \log \Delta f$$

where NL_0 is the marine environmental noise level at 1 kHz; NL_0 and constant c can be used to check Knudson marine environmental noise curve (see Fig. 2.9 in 2.3); The last term indicates that the output noise power of the receiver is proportional to the bandwidth of the system; γ_0 is the directivity aggregation coefficient of the receiving array at 1 kHz; The third and fourth terms on the right both describe the aggregation coefficient of conventional array at frequency f, which is proportional to the square of frequency. The directional characteristics of marine environmental noise are not very obvious, that is to say, the noise intensity incident from all directions to the receiving point is almost the same. If the receiving array has directivity, it mainly receives the noise within a solid angle in one direction, and the noise incident in other directions will be suppressed. Compared with the non-directivity receiving hydrophone, the output interference power of the directivity array is much smaller. The ratio of the two is described by the aggregation coefficient. According to acoustic theories, we get the following equation:

$$\gamma \cong \frac{4\pi A}{\lambda^2} = \gamma_0 f^2$$

where, A represents the cross-sectional area of the array.

The sound source level can be expressed as:

$$SL = SL_0 - 10b \log f + 10 \log \Delta f$$

where, $b = 6/3$ is the radiated noise level of the ship at 1 kHz. SL_0 and b can be determined by the experiment of ship noise. The above formula shows that the frequency dependence of ship radiated noise spectrum is a b negative power of frequency.

By substituting the above formulas into Eq. (1.25), we can get the following results:

$$60 + 15 \log r + 0.036 f^{3/2} = GL_0 + (2 + c - b) 10 \log f$$

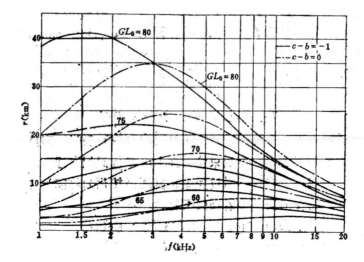

Fig. 1.9 Relationship between operating range of passive sonar and operating frequency and quality factor

$$GL_0 = -DT - NL_0 + SL_0 + 10 \log \gamma_0$$

The above equation shows that for a given high quality factor, the operating distance is closely related to the selection of working frequency. When designing sonar system, the above equation can be used to guide the selection of operating frequency, or for the required operating distance, the reasonable selection of f can only require a lower high quality factor, so the system is easier to realize or the cost of realizing the system is lower. It can be solved by graphic method or iterative method on computer, and the results are shown in Fig. 1.9 for the given case.

Figure 1.9 shows that the electric-acoustic performance of sonar is improved when the quality factor increases. For example, the size of array increases, the detection method is improved, so the detection threshold is reduced, or the noise level is reduced, etc.; the operating distance of sonar increases; The higher the quality factor is, the lower the working frequency should be chosen. In other words, the better the electric-acoustic performance of sonar is, the lower the working frequency should be; Only when the sonar works in low frequency and has good quality factor, can it have a larger working distance. Even if the high frequency sonar has good electro acoustic performance, it can not have a long detection distance. The reason is that the sound absorption of high frequency sound wave in seawater is larger, which is the fundamental reason for the development of modern long-range sonar to low frequency direction. The working frequency of modern towed linear array sonar is about 250 Hz. Figure 1.9 shows that when the working frequency is 20 kHz, the quality factor increases by 10 dB, and the increased operating distance will not exceed 1 km; When the working frequency is 5 kHz and the quality factor is increased from

70 to 10 dB, the operating distance will be increased by 14 km, which directly shows the reason why the long-range sonar must develop to low frequency.

Chapter 2
Average Energy Channel

The ocean is regarded as an acoustic channel, which has two main effects on sonar system: One is the mode of sound propagation in the ocean and the average loss of energy; The other is the transformation of the signal. The deterministic transformation leads to the distortion of the received waveform, and the random transformation leads to the loss of information. This chapter will discuss the first problem, which is to understand the conditions under which sound can travel further in the ocean. The second problem will be discussed in the following chapters.

2.1 Sound Velocity in Seawater

Sound velocity is the most important acoustic parameter in seawater, which has an important influence on sound propagation. Generally speaking, the sound velocity refers to the phase velocity of plane wave, and the average sound velocity in the ocean is about 1500 m/s. The speed of sound in seawater increases with the increase of seawater temperature, salinity and depth. The empirical formula [1] between them is as follows:

$$c = 1449.2 + 4.6T - 0.055T^2 + 0.00029T^3$$
$$+ (1.34 - 0.010T)(s - 35) + 0.016Z \tag{2.1}$$

where c is sound velocity, m/s; T means temperature, °C;
s refers to Salinity, ‰; Z means Depth, m.

On the sea, the relationship between temperature and water depth can be measured on the spot by using "automatic temperature depth recorder" (BT recorder), which is called temperature profile or temperature distribution. The salinity of seawater samples collected at different depths can be determined by chemical method or pH value. There are two types of BT recorders: the suspended type and the expendable

© Harbin Engineering University Press 2022
J. Hui and X. Sheng, *Underwater Acoustic Channel*,
https://doi.org/10.1007/978-981-19-0774-6_2

Fig. 2.1 Principle of
reverberation sound
velocimeter

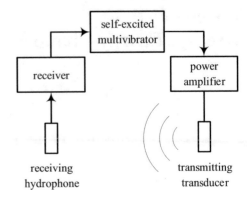

type. The former can be used repeatedly when the ship is drifting, and the latter can be used once when the ship is sailing. After obtaining the data of temperature, salinity and depth, the acoustic velocity can be calculated according to Eq. (2.1).

The sound velocity can also be measured on site by using the "reverberation" sound velocity meter [2], and its principle is shown in Fig. 2.1. The oscillation period of the multivibrator is determined by the propagation time of the sound wave from the transmitting transducer to the hydrophone, so the sound velocity is directly proportional to the frequency of the multivibrator, and the sound velocity can be obtained by measuring the frequency. There are also two types of acoustic velocimeters: the hanging type and the consumption type.

The distribution of sound velocity in the ocean is variable, including regional variation, seasonal variation, diurnal variation, and there is about 1 ‰ magnitude of variation in the near surface water layer within 2 h. An example of sound velocity distribution in the Atlantic Ocean is shown in Fig. 2.2, and that of sound velocity distribution in the water layer near the sea surface within 24 h is shown in Fig. 2.3. The unit of sound velocity in the figure is m/s.

2.2 Sound Absorption in Seawater

Seawater medium is lossy. In the process of propagation, sound energy is gradually lost and transformed into heat energy, which is called absorption loss. The absorption loss is related to seawater composition, temperature, pressure, frequency and propagation mode of sound wave. The absorption below 100 kHz is mainly due to the relaxation absorption of magnesium sulfate. Over 100 kHz, it is mainly because of the additional absorption caused by the viscosity of the medium. Below 5 kHz, the absorption loss is much larger than the relaxation absorption of magnesium sulfate, and the mechanism is the relaxation absorption of borate.

The empirical formula of absorption loss [1] is as follows:

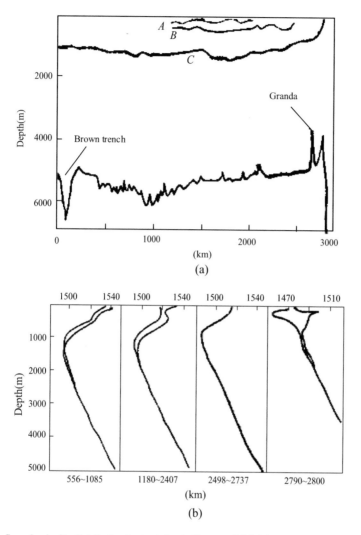

Fig. 2.2 Sound velocity distribution in the Atlantic Ocean. **a** Mid Atlantic sound velocity profile: *A*—shallow water channel axis; *B*—the lower bound of the surface channel; *C*—deep sea channel axis. **b** The distribution of sound velocity at different distances in the Atlantic Ocean is given by two sound velocity distributions

$$a = \frac{1.71 \times 10^8 (4\mu_F/3 + \mu_F') f^2}{\rho_F C_F^3} + (\frac{SA' f_{rm} f^2}{f^2 + f_{rm}^2})(1 - 1.23 \times 10^3 P)$$

$$+ \frac{A'' f_{rb} f^2}{f^2 + f_{rb}^2} \quad (dB/m) \tag{2.2}$$

where, $\rho_F \simeq 1000 \, \text{kg/m}^3$—seawater density;

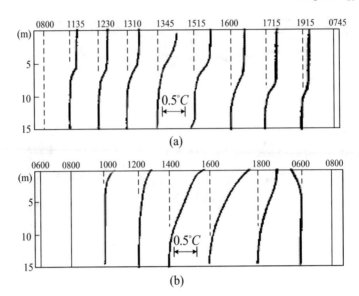

Fig. 2.3 Diurnal variation of temperature distribution. **a** The wind speed is fast enough to be 3–8 m/s. Due to the agitation of sea water, the isothermal layer always appears. **b** When the wind speed is low, the sea water becomes calm, and the isothermal layer appears; Only in the early morning and late at night; In the daytime, especially in the afternoon, there is a negative gradient water layer with a wind speed of 2 m/s

$C_F \simeq 1461$ m/s—Sound velocity (salinity is 0, and the temperature is 14 °C);

$\mu_F \simeq 1.2 \times 10^{-3}$ N s/m² (= 14 °C)—dynamic shear viscosity coefficient of fresh water;

$\mu_F' \simeq 3.3 \times 10^{-3}$ N s/m² (= 14 °C)—dynamic volume viscosity coefficient of fresh water;

$f_{rm} = 21.9 \times 10^{[6-1520/(T+273)]}$, kHz; Relaxation frequency of magnesium sulfate.

$f_{rb} = 0.9 \times (1.5)^{T/18}$, kHz; Relaxation frequency of borates.

$A' = 2.03 \times 10^{-5}$, dB/(kHz)(m)(10⁻³);

$A'' = 1.2 \times 10^{-4}$, dB/(kHz)(m);

S—Salinity, ‰;

f—Acoustic frequency, kHz;

P—Static water pressure, Pa.

The temperature dependence of relaxation frequency of magnesium sulfate and borate is shown in Fig. 2.4 and Fig. 2.5 respectively.

The relationship between the absorption coefficient and frequency of seawater near the sea surface ($T = 14$ °C, $S = 35$ ‰) calculated by Eq. (2.2) is shown in Fig. 2.6 and Fig. 2.7. The absorption loss of high frequency sound wave is considerably large. During World War II, the frequency of active sonar was about 30 kHz, and the absorption loss per kilometer was about 6 dB. Even if the system is improved to increase the quality factor by 12 dB (for example, the detection threshold is reduced

Fig. 2.4 Relationship between relaxation frequency f_{rm} and temperature T of magnesium sulfate

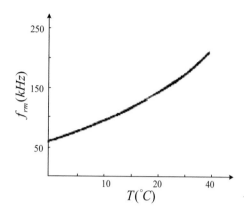

Fig. 2.5 Relationship between relaxation frequency f_{rb} and temperature T of borate

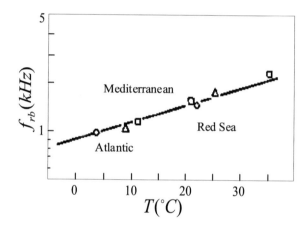

by 12 dB), the effect can only increase the operating distance by less than 1 km. When the working frequency of modern sonar is 3 kHz or lower and the sonar is also improved by 12 dB, the operating range will be doubled. The decrease of absorption loss with frequency is one of the main reasons for the development of modern sonar to low frequency. The dynamic shear and volume viscosity coefficients of fresh water are shown in Fig. 2.8.

2.3 Marine Environmental Noise

Another important acoustic property of marine media is marine environmental noise, which has an important impact on the detection and recognition of sonar system, as one of the main background interferences. According to the causes of noise, the marine environmental noise can be divided into:

Fig. 2.6 Absorption loss in
high-frequency section

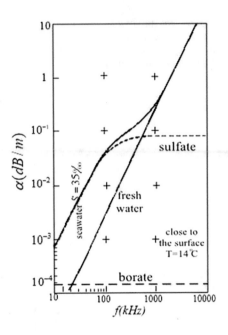

Fig. 2.7 Absorption loss in
low-frequency section

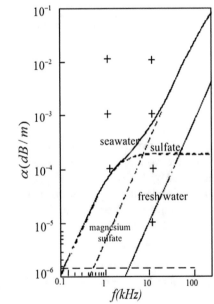

Fig. 2.8 Dynamic shear and volume viscosity coefficient of fresh water

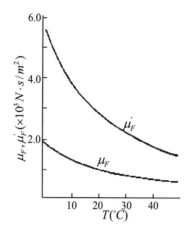

Marine dynamic noise: it is related to wind and waves, and it is the noise produced by turbulence in sea water and atmosphere. It also includes the noise caused by waves, rain, bubbles, etc.

Bio noise: the sound produced by various organisms.

Traffic noise and industrial noise: it is caused by human activities.

Seismic noise: noise produced by earthquakes, volcanism and tsunamis.

Subglacial noise: noise caused by the formation and movement of ice.

Marine environmental noise is complex and variable, which is related to the location of the sea area, the position of the hydrophone, the meteorological conditions in the near and far areas, as well as the frequency.

When it is below 20 Hz, the main noise sources are ocean turbulence, earthquake and tide; When it is between 20 and 500 Hz, the main noise is traffic noise; When it is above 500 Hz, the main noise sources are sea waves and their broken waves; Above 50 kHz, the thermal noise is mainly caused by the motion of seawater molecules.

Knudson et al. [3] studied the spectral level of marine environmental noise in 1948. The so-called spectral level of environmental noise refers to the sound pressure level within 1 Hz bandwidth received by non-directional hydrophone. Their research results show that the ambient noise level increases with the sea state or wind force, and decreases by 5–6 dB per octave with the increase of frequency, as shown in Fig. 2.9.

Figure 2.10 shows the modern measurement results [4] of Wenz and other authors. The area A indicated by the shadow line is the result of seafloor measurement with seismometer in the Pacific Ocean; The results measured in Trieste deep diving are represented by the shadow area B; The shaded areas C represent the results measured in inland deep lakes; Wenz's measurement results are represented by thick solid lines D; The curve E is raindrop noise; The straight line F is thermal noise.

The marine environmental noise follows the law of Gaussian distribution, i.e., the probability density of the instantaneous value x of the sound pressure is as follows:

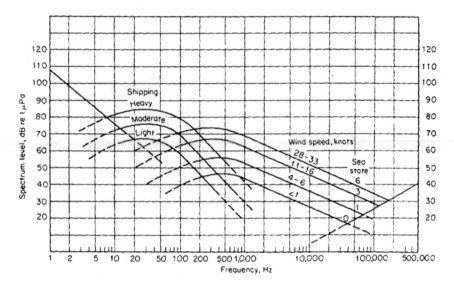

Fig. 2.9 Spectrum level of deep-sea environmental noise

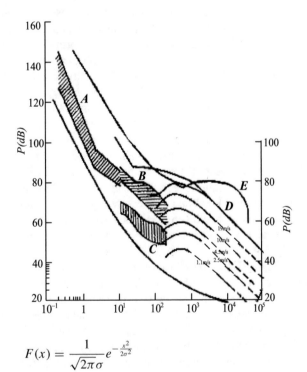

Fig. 2.10 Marine environmental noise

$$F(x) = \frac{1}{\sqrt{2\pi}\sigma} e^{-\frac{x^2}{2\sigma^2}}$$

where σ—the standard deviations of x.

σ^2—the mean square strength of noise.

The noise spectrum of marine environment in Fig. 2.9 is measured by non-directional hydrophone. Delicate experiments show that: in fact, marine environmental noise has both vertical distribution and horizontal distribution. That is to say, the ambient noise is not isotropic, and the noise intensity of different directions reaching the receiving point is quite different. The noise above 1 kHz mainly comes from the sea surface. Compared with the horizontal direction, the noise level from the sea is 5–7 dB higher; For the noise under 1 kHz, especially at 100 Hz, the level from the horizontal direction is 20 dB higher than that from the water surface, which indicates that the main noise source in this frequency band is traffic noise.

Figure 2.11 is an example of marine environmental noise with vertical distribution, and the measurement is completed in the deep sea.

For hundreds of Hz environmental noise, the main noise source is traffic noise, so it is usually variable and has obvious horizontal directionality. Figure 2.12 shows the example of traffic noise with horizontal distribution.

Fig. 2.11 Vertical distribution of environmental noise in deep sea

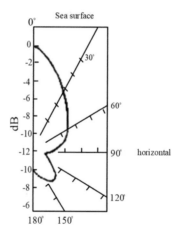

Fig. 2.12 Example traffic noise with horizontal distribution

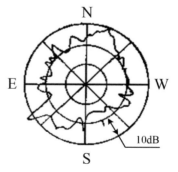

The spatial correlation radius of single frequency marine environmental noise field is less than half a wavelength, especially in the vertical direction. The spatiotemporal statistical characteristics of environmental noise field are not detailed here, and readers can refer to the relevant special literature.

Using Fig. 2.9, under the condition of narrow band, The center frequency can be easily estimated as f_0 and the bandwidth can be the noise level in the passband of Δf by using the following calculation formula:

$$NL_{\Delta f} = NL(f_0) + 10 \log \Delta f$$

where $NL(f_0)$ can be seen from Fig. 2.9.

2.4 Seabed Reflection Loss

The acoustic characteristics of seabed interface have an important influence on sound propagation, especially in shallow water. In fact, the seabed is uneven, and it is assumed that the bottom interface is flat, but the results still have many applications. In fact, the seabed is not liquid, but the medium is assumed to be liquid, and the theoretical results are still in reasonable agreement with the experimental results.

The reflection coefficient of plane sound wave projected on the interface of two-layer liquid at a certain grazing angle (the angle between the surface and the horizontal plane) was first derived by Rayleigh. If the sweep angle between the projected sound wave and the interface is θ_1, as shown in Fig. 2.13, then the sweep angle between the reflected sound line and the interface is also θ_1. Set the density of the two-layer liquid to be ρ_1 and ρ_2 respectively, the sound velocity to be c_1 and c_2 respectively, and the sound intensity of the projected wave and the reflected wave to be I_i and I_r respectively, we have [5]:

$$\frac{c_1}{\cos \theta_1} = \frac{c_2}{\cos \theta_2} \text{——(Law of refraction)}$$

Fig. 2.13 Acoustic reflection and refraction at the interface of two-layer medium

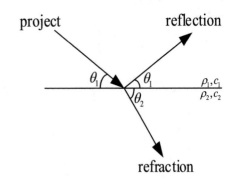

$$\frac{I_r}{I_i} = \left[\frac{m\sin\theta_1 - n\sin\theta_2}{m\sin\theta_1 + n\sin\theta_2}\right]^2 = \left[\frac{m\sin\theta_1 - (n^2 - \cos^2\theta_1)^{1/2}}{m\sin\theta_1 + (n^2 - \cos^2\theta_1)^{1/2}}\right]^2$$

where, $m = \frac{\rho_2}{\rho_1}, n = \frac{c_1}{c_2}$.

The reflection coefficient of seabed is the value of the above formula, and the seabed reflection loss is the decibels of the above formula, that is $10\log\frac{I_r}{I_i}$ (dB). Therefore, the reflection loss as a function of the sweep angle of the projected sound wave is related to the ratio sum m and n. For all four different cases, its characteristics are shown in Fig. 2.14 [4]. Most of the seabed in the ocean is shown in Fig. 2.14c, that is, there is a critical angle of total reflection/omni-reflection. When the grazing angle is less than the critical angle of total reflection, total acoustic reflection appears, and the reflection loss will be zero. For the muddy seabed, where the sound velocity is less than that in the sea, the reflection is shown in Fig. 2.14a.

The above discussion does not consider the sound absorption of seabed medium, which actually has a large sound absorption. The absorption effect will not make the reflection loss be zero when the incident sound wave is less than the critical angle of total reflection, and the reflection coefficient image depending on the grazing angle becomes smoother. The dotted line in Fig. 2.14c turns out to be the reflection coefficient after considering the sound absorption of seabed medium.

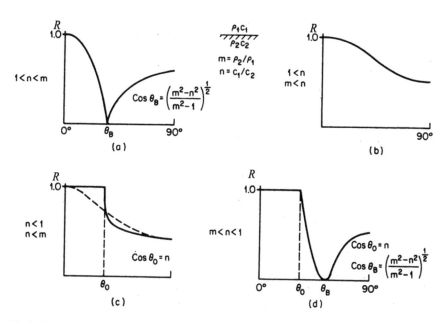

Fig. 2.14   Seabed acoustic reflection coefficient **a** $1 < n < m$; **b** $1 < n, m < n$; **c** $n < 1, n < m$; **d** $m < n < 1$

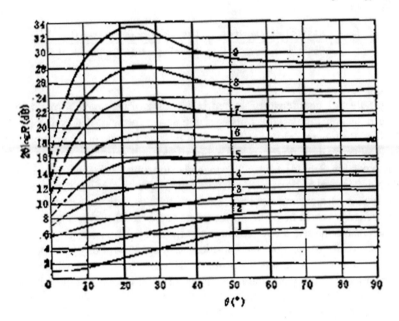

Fig. 2.15 Seabed reflection loss

Figure 2.15 shows the angle dependence of seabed reflection loss [4], which is the result of semi experience and semi theory. The reflection loss of a particular sea area can always be fitted with one of the curves.

2.5 Ray Acoustic in Layered Media

Two methods are used to study sound propagation in the ocean: wave acoustics and ray acoustics. This section mainly describes ray acoustics. The concept of ray originated from people's intuitive sense of the straightness of light. Many problems in optics can be easily explained with the concept of ray. Ray optics method is widely used in optics. Acoustics borrows the concept of ray. Compared with the wave method of sound, the theory of ray has the advantages of intuition and simplicity.

The basic assumption of ray acoustics is that sound energy is transmitted along a certain direction line, which is called sound line, and the sound line is perpendicular to the wave front.

According to the assumption, the sound energy is transmitted along the sound line, so the sound energy will not pass through the wall of the tube.

If the acoustic tube is thin enough, the sound intensity at each point in the same section can be regarded as equal. In lossless medium, the given sound energy transmits along the acoustic tube, and the sound intensity at different sections needs to be inversely proportional to the cross-sectional area (the sound intensity is the sound energy passing through the unit area). The phase difference of sound wave in two different sections of the same acoustic wiring harness depends on the distance along the sound line between the two sections and the sound velocity at this position. The total phase change of sound wave from point M to point N along the sound line is observed. The sound segment is divided into many small segments ds. The phase passing through ds appears to be $\frac{2\pi ds}{\lambda}$, and the total phase is changed to be $\int_{\widehat{MN}} kds$ along the sound line from M to N. $k = \frac{2\pi}{\lambda}$ is called wave number. In ray acoustics, in order to calculate the phase change conveniently, the concept of sound path $L(\mathbf{r})$ from point M to point N along the sound ray is defined as:

$$k_0 L(\mathbf{r}) = \int_{\widehat{MN}} kds$$

Or it is changed to a differential form:

$$dL(\mathbf{r}) = \frac{kds}{k_0}$$

i.e.

$$\frac{dL}{ds} = \frac{k}{k_0} \tag{2.3}$$

In Eq. (2.3), it is the wave number k at a point on the sound line, and it is the wave number k_0 at a selected reference point. Therefore, the sound path $L(\mathbf{r})$ from point M to point N along the sound line is the distance that the sound wave passes through in the uniform medium with wave number k_0. When passing through the distance, the phase change of the sound wave is equal to the phase change along the sound line $\widehat{MN}$ in the actual medium.

The surface caused by the point with equal sound range $L(\mathbf{r})$ is equal phase plane, which is called wave array surface, so the wave array surface is also called equal acoustic range surface. According to the basic assumption of ray acoustics, the sound line should be perpendicular to the wave front, so the derivative along the sound line in Eq. (2.3) can be written as a gradient. Accordingly, Eq. (2.3) can be rewritten as follows:

$$(\nabla L)^2 = n^2, n = k/k_0 \tag{2.4}$$

where, n denotes the Acoustic refraction coefficient.

Equation (2.4) is called sound path equation, which is the basic relation of ray acoustics. In fact, Eq. (2.4) can also be put forward as the basic hypothesis of ray acoustics.

In essence, sound wave is a wave phenomenon, and sound line is an approximate concept. Here we omit the proof and point out that ray acoustics is applicable in two cases: the first case is when the sound field is not very different from spherical wave and plane wave, or the spatial change of sound field is not very sharp; The second case, the sound field of high frequency sound wave, is reasonable for the approximation of high frequency sound wave rays.

From the previous section, we know that when plane wave passes through the interface of two kinds of fluid medium, reflection and refraction will occur. The relationship between incidence angle and refraction angle (see Fig. 2.13) satisfies the refraction law:

$$\frac{c_1}{\cos\theta_1} = \frac{c_2}{\cos\theta_2} \tag{2.5}$$

When the sound velocity of medium changes continuously with a coordinate, what form will the refraction law take? Let's generalize Eq. (2.5). Suppose that the medium is divided into many layers, each layer is very thin, then the sound velocity in each layer can be regarded as even. When Eq. (2.5) is applied to each interface, it will be:

$$\frac{c_1}{\cos\theta_1} = \frac{c_2}{\cos\theta_2} = \dots = \frac{c_i}{\cos\theta_i} = \dots = \text{constant}$$

c_i is the sound velocity at layer i, θ_i is the grazing angle at the boundary of layer i. If the sound velocity at a reference point is c_0 and the sweep angle of the sound line at this point is θ_0, the refraction law in layered media is as follows:

$$\frac{c_i}{\cos\theta_i} = \frac{c(z)}{\cos\theta(z)} = \frac{c_0}{\cos\theta_0} \tag{2.6}$$

Equation (2.6) assumes that the coordinate axis z is perpendicular to the layered interface.

In the literature, the general Eq. (2.6) is Snell law. Its extension process is not strict, which cannot be referred to as proof, but a kind of popular narration. Snell law shows that the path of sound line will be a curve in layered medium with continuous change of sound velocity. The formula for calculating the track of sound line is given below.

When the horizontal coordinate is r, the z coordinate is vertical and pointing downward, the sound velocity of the medium is a function of z. See Fig. 2.16 for a section of micro unit on the sound line

Fig. 2.16 Micro unit of sound line

$$dr = dz/tg\theta$$

i.e.

$$r = \int \frac{dz}{tg\theta}$$

The positive direction of sweep angle θ is clockwise.

If the sound source is located at (r_0, z_0), considering Snell's law and taking θ_0 as the initial grazing angle, any point (r, z) on the sound line emitted from the sound source should satisfy the following sound line trajectory equation:

$$r - r_0 = \int_{z_0}^{z} \frac{dz}{tg\theta} = \int_{z_0}^{z} \frac{\cos\theta_0}{\sqrt{n^2(z) - \cos^2\theta_0}} dz \tag{2.7}$$

where, $n(z) = \frac{k(z)}{k_0} = \frac{c_0}{c(z)}$.

For a given sound velocity distribution function $c(z)$, the sound ray trajectory can be obtained from Eq. (2.7).

The time required for passing through a ray unit ds is $dt = ds/c(z)$. As can be seen from Fig. 2.16, $ds = \cos\theta dr + \sin\theta dz = dz/\sin\theta$, considering Snell's law, the time required for sound wave to transmit from (r_0, z_0) to (r, z) along the sound line is t:

$$t = \frac{\cos\theta_0}{c_0}(r - r_0) + \frac{1}{c_0} \int_{z_0}^{z} \sqrt{n^2(z) - \cos^2\theta_0} dz$$
$$= \frac{1}{c_0} \int_{z_0}^{z} \frac{n^2(z)dz}{\sqrt{n^2(z) - \cos^2\theta_0}} \tag{2.8}$$

2.6 Sound Lines and Fields in Constant Gradient Water Layer

The decrease of sound velocity with the increase of sea depth is called negative gradient distribution, otherwise it is called positive gradient distribution.

When the ocean is calm and the sunshine is strong, the surface of the ocean is heated, but the water temperature in the deep water is not, which leads to a negative sound velocity gradient distribution. The strong wind wave will mix the surface of the ocean evenly, and an isothermal water layer with a certain thickness will appear. Because the sound velocity increases with the increase of the static pressure of the sea water, the isothermal water layer is a positive sound velocity gradient water layer.

For the constant gradient water layer, the function form of sound velocity distribution is as follows:

$$c(z) = c_0[1 + a(z - z_0)] \tag{2.9}$$

When $a > 0$, it is positive gradient distribution, and when $a < 0$, it is negative gradient distribution. a is called relative sound velocity gradient.

Figure 2.1 shows the sound ray diagram for the negative gradient distribution. For example, the position of the source is $(0, z_0)$, the sound velocity of the depth of the source is c_0, thus substituting Eq. (2.9) into Eq. (2.7), the sound ray trajectory equation can be obtained by

$$r = \frac{1}{a \cos \theta_0} (\sin \theta_0 - \sin \theta) \tag{2.10}$$

where θ_0—the sweep angle of sound line leaving sound source; θ—the sweep angle at a point on the sound line.

Substituting Eq. (2.9) into Eq. (2.6), we can see that:

$$z - z_0 = \frac{1}{a \cos \theta_0} (\cos \theta - \cos \theta_0) \tag{2.11}$$

It is not difficult to eliminate θ from Eqs. (2.10) and (2.11)

$$\left(r - \frac{1}{a} tg\theta_0\right)^2 + \left(z - z_0 + \frac{1}{a}\right)^2 = \frac{1}{a^2 \cos^2 \theta_0} \tag{2.12}$$

The trajectory equation of sound ray in Eq. (2.12) is a circular equation. It can be seen that all sound lines in the water layer with equal gradient are arc lines with radius of $1/a \cos \theta_0$. The difference between the negative gradient water layer and the positive gradient water layer is that the sound line of the former bends downward; In the latter case, the sound line bends upward, which can be summarized in one sentence, that is, the sound line tends to bend to the water layer with low sound speed, which is called the refraction effect.

The sound ray shows the distribution of sound field intuitively. For example, the sound field in the seawater layer with negative gradient sound velocity distribution shown in Fig. 2.17 is visually divided into "bright area" and "shadow area". Due to the refraction effect, the upward sound line emitted by the sound source at a small grazing angle bends to the depth of the sea before reaching the sea surface. At the

Fig. 2.17 Sound ray in negative gradient distribution

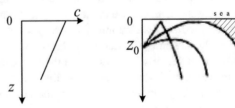

depth where the sound line reverses, there should be $\theta = 0$, and the minimum depth z_{min} can be obtained from formula (2.11) as follows:

$$z_{min} = z_0 + \frac{1}{a\cos\theta_0}(1 - \cos\theta_0), \quad a < 0 \tag{2.13}$$

The sound line tangent to the sea surface is called the limit sound line. The left side of the limit sound line is called the "bright area". Any point in the bright area has a direct sound line passing through. The right side of the limit sound line is the "shadow area". No direct sound line can enter the shadow area. From the perspective of ray acoustics, the sound intensity in the shadow area is zero.

The sweep angle of the limit sound line at the sound source is as follows:

$$\theta_L = -\arccos\frac{1}{1 - az_0}$$

The boundary of shadow area is:

$$r_L = \sqrt{\frac{2z_0}{a}} + \sqrt{\frac{2z}{a}}$$

The order of magnitude of a is about 1×10^{-4} 1/m. The curvature radius of the sound line is $1/a\cos\theta_0$, and the minimum curvature radius is about 10 km, so the sound line bending is actually small, but this small sound line bending has a significant impact on the sound field, resulting in a shadow area.

In order to deepen the concept that the slight bending of sound line will have an important influence on the sound field, we need to further study the example in Fig. 2.18.

In the surface layer of the ocean, due to the disturbance of wind and waves, there is a certain thickness of isothermal layer, which is adjacent to the negative gradient water layer. The isothermal layer with positive sound velocity gradient on the surface is called surface channel, and its relative sound velocity gradient is 10^{-5} 1/m, its thickness is between 20 and 150 m. In the negative gradient layer, the relative sound gradient is larger, which is about 10^{-4} 1/m.

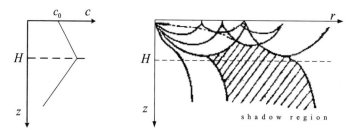

Fig. 2.18 Sound ray in surface channel

The sound velocity distribution shown in Fig. 2.18 can be expressed as a function:

$$c(z) = \begin{cases} c_0(1+az), & 0 \le z \le H \\ c_0(1+aH)[1-b(z-H)], & H \le z \end{cases}$$

If the position of the sound source is taken as $(0, 0)$, the sound track emitted from the source at θ_0 angle can be obtained from Eq. (2.10):

$$r = \begin{cases} \dfrac{1}{a\cos\theta_0}(\sin\theta_0 - \sin\theta), & 0 \le z \le H \\ \dfrac{1}{a\cos\theta_0}(\sin\theta_0 - \sin\theta_H) + \dfrac{1}{b\cos\theta_H}(\sin\theta - \sin\theta_H), & H < z \end{cases}$$

where θ_H—the sweep angle of sound line at depth H.

The relationship between θ_H and θ_0 is related by Snell's law:

$$\sin\theta_H = \sqrt{1 - (1+aH)^2 \cos^2\theta_0}$$

It has been pointed out that in the positive gradient water layer, the sound line is an arc bending to the sea surface, and in the negative gradient water layer, it is an arc bending downward. The sound line tangent to the isobath $z = H$ is called the critical sound line, and its exit angle θ_c from the sound source is recorded, which divides all the sound lines into two types. At a certain depth less than H, the sound line of $\theta_0 < \theta_c$ is reversed and deflected to the sea surface, and transmits in the positive gradient water layer forever after being reflected by the sea surface. The sound lines of $\theta_0 > \theta_c$ bend downward through the negative gradient layer, and they will not return to the isothermal layer without reflection. The critical sound line splits into two branches at the inversion depth $z = H$, one of which bends to the sea surface; Second, it bends down to form a shadow area on the right side of the split (as shown by the shadow line), where there is no direct sound line passing through.

Now let's talk about the sound field in the isothermal layer. The sound lines of $\theta_0 < \theta_c$ contribute to the sound field in the isothermal layer, and the maximum depth they reach is known by Snell's law:

$$z_{\max} = \frac{1 - \cos\theta_0}{a\cos\theta_0}, \quad \theta_0 < \theta_c$$

Under the condition of $\theta_0 \ll 1$, we get $\cos\theta_0 \approx 1 - \frac{\theta_0^2}{2}$. Substituting into the above formula, we obtain:

$$z_{\max} = \frac{\theta_0^2}{2a}$$

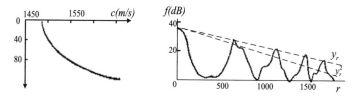

Fig. 2.19 Results of surface channel sound field experiment (simulation experiment)

It is not difficult to see that this kind of sound line leads to the concentration of sound energy near the sea surface. The meaning of the above formula can be explained as that the sound lines from the sound source with grazing angle between 0 and θ_0 are concentrated in the water layer with thickness of $\theta_0^2/2a$. The thickness of this layer decreases with the square law of θ_0, while the sound energy emitted to this layer decreases with the power of θ_0. It can be seen that the smaller the θ_0 is, the higher the sound energy concentration will be. Thus, a high sound intensity area appears near the sea surface.

There is an envelope surface of sound line in the isothermal water layer, as shown by the dotted line in Fig. 2.18. The cross-sectional area of the sound beam tube at the envelope surface is zero. The cross-sectional area of the sound beam tube from the sound source expands gradually and converges again at the envelope surface to form a high-intensity area, which is called "convergence area". It is a unique phenomenon of the sound channel transmission mode that high-intensity convergence areas appear alternately.

Figure 2.19 shows the relationship between the sound intensity in the surface channel and the horizontal distance measured in the flume simulation experiment, and the experimental results are in good agreement with the theoretical prediction. In the surface channel, the average attenuation rate of sound intensity with distance is slower than that of spherical wave, but close to that of cylindrical wave (which is also called $\frac{1}{r}$).

Both theory and practice have proved that the sound wave can travel far in the surface channel. When the isothermal layer is thick enough and the sea surface wave is small, there will be obvious surface channel effect.

In the northern China Sea area, the isothermal layer can extend to the sea bed in winter. In this case, the positive gradient layer is more stable and the sound transmission condition is better.

In the surface layer of sea water, a large number of bubbles are produced due to wind and waves. The absorption and scattering of bubbles will produce additional attenuation of sound intensity. The sea surface is not flat, and the reflection of sound waves on the uneven sea surface will also lead to the loss of sound intensity in the direction of sound propagation.

Figure 2.20 shows the experimental results of the sound field in the surface acoustic channel of the ocean. The frequency of the sound wave is 1030 Hz and the thickness of the isothermal layer is 100 m. The black spot represents the relative sound intensity measured experimentally. The results of the modern modified ray theory and wave

Fig. 2.20 Attenuation law
of sound intensity in the
surface channel of the ocean

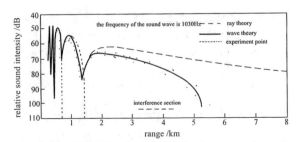

theory are also shown in the figure. The theoretical and experimental results are
in good agreement, especially the wave theory. The experimental point has a large
dispersion range, which is caused by the up and down swing of the sound source
and hydrophone and the random wave of the water surface, so the amplitude of the
received signal is not stable, which is called signal fluctuation.

2.7   Deep Sea Sound Transmission Mode and Spreading
Loss

Not only the distribution of sound velocity in the ocean, but also seabed and sea
surface are the main factors affecting sound propagation. A typical deep-sea sound
velocity distribution is shown in Fig. 2.21. The sound velocity distribution can be
divided into four layers. The surface layer has obvious diurnal variation, which is
sensitive to sunshine, temperature and wind waves. The surface layer is usually an
isothermal layer with a thickness of about 30–100 m. It forms the so-called surface
sound channel, and has good sound transmission conditions when the layer thickness
is large. After a long period of remaining calm, the isothermal layer will disappear.

Fig. 2.21 Typical
distribution of sound velocity
in deep sea

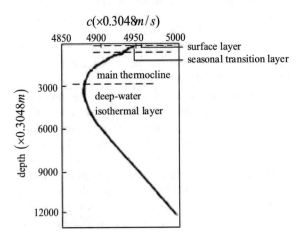

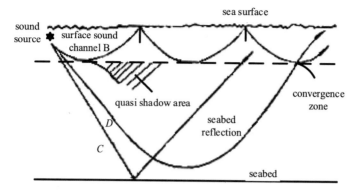

Fig. 2.22 Schematic diagram of three transmission modes in deep sea

The depth of the water layer below the surface layer is less than 300 m, which is called thermocline. The sound velocity decreases sharply with the increase of depth. This layer is affected by the season. There is a minimum sound velocity point at the depth of 800–1200 m, which is called the channel axis. The water layer with negative sound velocity gradient above the sound channel axis is called the main thermocline, and the isothermal water layer with water temperature about 2 °C below it is the water layer with positive sound velocity gradient. The main thermocline and deep-water isothermal layer form the so-called deep-sea channel (SOFA channel). The explosion sound of several kilograms of TNT equivalent can reach several thousand kilometers near the sound channel axis.

In radar and radio communication, there are so-called ground wave, sky wave and scattered wave. Similarly, the sonar working in the deep sea near the sea surface uses three different transmission modes, as shown in Fig. 2.22. Sonar designers should choose one or more modes of transmission in advance, and make sonar system adapt to the selected mode of transmission.

The research on the three modes of communication promoted the development of sonar in the 1960s, and made the sonar's operating distance leap for the first time after the war.

Before 1960, the ship hull sonar in the other countries basically only used the surface channel to work. Due to the static pressure of sea water, the surface isothermal layer forms a positive sound velocity gradient layer. The sound energy emitted from the sound source at a small grazing angle enters the surface channel, which is basically retained in the channel due to the refraction effect. Only part of the sound energy is leaked due to the scattering of the uneven sea surface and the transverse diffusion of the energy at the lower boundary of the surface layer along the wave front. There is a "cut-off frequency" in the surface channel. Below this frequency, the channel appears too "thin" to trap the acoustic energy. The degree of acoustic energy leakage from the waveguide will increase rapidly with the decrease of frequency, and the expansion loss will increase greatly. The thicker the surface layer is, the lower the cut-off frequency is. The cut-off frequency is about 1000 Hz for 30 m isothermal layer.

Therefore, it is not suitable to use too low frequency for sonar using surface channel working mode; When the working frequency is too high, the scattering leakage caused by uneven sea surface will increase, which proves to be inappropriate.

Under the surface layer, the area where the direct sound cannot reach (the shadow line area in the figure) is the so-called "quasi shadow area". Both the sound energy leaked from the surface channel and the sound reflected from the seabed enter the area, and the sound intensity in the quasi shadow area is very small. When the sonar is used to search the target in the shadow area, its operating distance is less than 5 km.

Sonar makes use of seabed reflection and convergence area transmission mode, so that the range of sonar is greatly improved. Figure 2.23 shows the typical transmission loss of the three modes of transmission. It can be seen that at a distance of 10n mile, the reflected sound from the bottom of the sea is much stronger than the direct sound from the surface channel, and it can "shine through the shadow area", so the sonar operating distance using the reflected sound from the bottom of the sea can be increased to more than 15n mile. In short distance, the angle of sound wave projecting to the seabed is large, and most of the sound energy is absorbed by the seabed or penetrated into the seabed medium. Those sound lines whose projection angle is less than the critical angle of total reflection of the seabed medium can make a major contribution to the sound field at medium distance, so that the sound intensity exceeds the direct sound arriving in the surface channel. In order to make use of the seabed reflection transmission mode, the sonar system must adopt a lower frequency of several kHz; The beam should be able to rotate to 45° in the vertical plane; The high quality factor of active sonar and passive sonar should be 185 dB and 100 dB

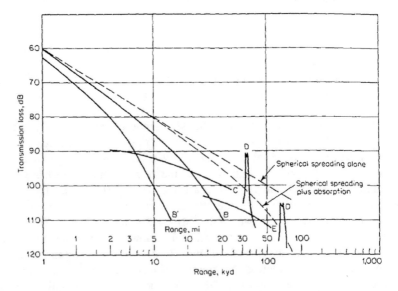

Fig. 2.23 Transmission loss of three modes in deep sea

respectively; In order to calculate the accurate horizontal distance of the target, the sonar system should be equipped with sound ray tracker and precision depth sounder.

If the sea depth is large enough, the sound velocity at the bottom of the sea is greater than or equal to the sound velocity on the sea surface. In this case, the sonar near the sea surface can make use of the convergence zone effect, and there is a searchable area about 3 km wide near 30n mile. The blind area of sonar is between the limit operating distance and the convergence area of the submarine reflection mode. When the sonar beam rotates to a certain angle range, the sound energy emitted from the sound source transmits along the refracted sound lines that neither touch the sea surface nor the sea bottom (see Fig. 2.22, sound line D). Due to the refraction effect, these sound lines converge near the sea surface again at 30–35n miles. The envelope of a cluster of sound lines is called caustics, and the high-intensity area near the collecting and distributing line is called convergence area. Experiments show that the sound intensity in the convergence region is 25 dB higher than that in the spherical wave pattern, and usually at least 10–15 dB higher.

Figures 2.24 and 2.25 provide measured examples of deep-sea transmission losses. The data in Fig. 2.24 are measured by explosion sound source, and the center frequency of filter is 100 Hz. It can be seen that a series of high-intensity convergence areas are gradually blurred at a distance of 350 km. By contrast, the spherical transmission loss curve is drawn with a thick line, which shows that the transmission loss at the convergence area in the deep sea is more than 10 dB smaller than that of the spherical transmission loss. Figure 2.25 shows the experimental results obtained from continuous low-frequency signals [6]. A series of high-intensity convergence regions can also be seen. Because of the use of sinusoidal radiation sound waves with stable frequency, the receiver can use a digital filter with narrow bandwidth to eliminate background interference, and the low-frequency sound radiation of hundreds of watts can be received thousands of kilometers away, which makes it possible for ultra long range underwater acoustic communication and detection.

Fig. 2.24 Experimental results of deep sea transmission loss

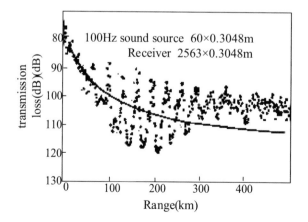

Fig. 2.25 Experimental results of deep sea CW ultra-long range transmission. The source depth of 111.1 Hz is 21 M; The depth of 13.89 Hz sound source is 110 m

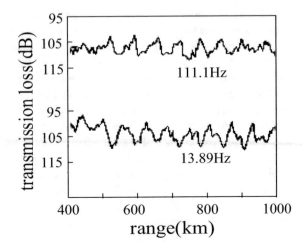

2.8 Pekeries Model in Shallow Water

Most of the sea areas in China are shallow waters, and the depth of the shallow waters on the continental shelf is within 200 m. The condition of sound transmission in shallow water is worse than that in deep sea. At present, the operating distance of Shipborne sonar in shallow water is not far. Only the towed linear array sonar working in low frequency band has a long operating distance, and its working frequency is about 250 Hz. For such a low frequency, the wave theory is more suitable to analyze the sound field in shallow water.

The characteristic of sound transmission in shallow water is that the influence of seabed is very serious.

From the perspective of engineering application, the first concern is not mathematical problems, but to understand the main physical factors and physical phenomena that affect the sound transmission in shallow water. This section follows the above principles and omits the complicated and tedious mathematical description.

The sound wave from the sound source transmits to the distance through many times of reflection from the seabed and sea surface. The uneven interface scatters the sound energy, which leads to the energy loss of the forward transmitting sound wave. Especially, the sound energy gradually leaks into the seabed or is absorbed by the seabed medium in the process of propagation. Therefore, only the sound line with small grazing angle plays an important role in the long-distance sound field. The vertical distribution of sound velocity affects the trajectory of the sound line, which affects the number of times that the sound line touches the seabed and the reflection loss of the seabed. Therefore, the main factors that determine the characteristics of sound transmission in shallow water are sea state, sediment, sea depth, sound velocity profile, sound frequency and the depth of transmitting transducer and receiving hydrophone.

For the sound waves with frequencies above 1 kHz, the influence of sound velocity distribution is important. In winter, the isothermal layer often appears in the northern sea area, and the sound transmission condition is similar to the surface channel. Under this condition, the sonar has a long operating distance. In some special sea areas, due to the existence of deep current, there will be a special vertical distribution of salinity or temperature, and a minimum of sound velocity at a certain depth, forming the so-called shallow underwater channel, which is a better sound transmission condition than the surface channel. In summer, especially in the southern sea area, the strong sunlight forms a large negative temperature gradient. Due to the refraction effect of the negative sound velocity gradient, the sound line bends sharply to the seabed, resulting in a great loss of sound transmission, which is often encountered in shallow water. In shallow water, a strong but very thin water layer with negative sound velocity gradient often appears at a certain depth, which is called the thermocline. The thermocline has a significant shielding effect on high-frequency sound waves, and the ability of high-frequency sonar to search the target opposite to the layer is significantly reduced. For tens to hundreds of Hz sound waves, the velocity distribution has little effect on the sound transmission in shallow water.

The rest of this section deals with low frequency sound transmission in shallow water. In order to meet the needs of engineers, the mathematical problem of wave theory is omitted, and the wave phenomenon of sound transmission is briefly introduced.

In 1944, Pekeries proposed the following shallow water model [7] in order to explain the experimental results of explosion pulse transmission in shallow water: the seabed and sea surface are two plane interfaces, where, the sea surface is absolutely soft in acoustics, the thickness water layer is H, the density of sea medium is ρ_1, and the sound velocity is c_1; It is assumed that the seabed medium is a liquid with a density of ρ_2 and a sound velocity of c_2. Generally speaking, $c_2 > c_1$. The Pekeries model seems too simple at first, and the seabed medium is not liquid in fact. However, the theoretical predictions based on the model are in good agreement with the experimental results, which proves the scientific authenticity of the model. It means that for the sound waves with frequencies higher than the seismic wave, only the P-wave acoustic parameters of the seabed and the layered structure of the seabed play a major role in the sound propagation. Figure 2.26 shows the Pekeries model in shallow water.

Fig. 2.26 Shallow water model

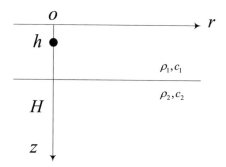

From the point of view of wave acoustics, the study of sound field in shallow water can be reduced to solve wave equations and find suitable boundary conditions. The sound field can be described as the sum of a series of normal modes

$$p(r, h, z) = \pi \omega \rho_1 \sum_n U_n(z) U_n(h) H_0^{(1)}(k_n r)$$

$$= \pi \omega \rho_1 \sum_n \varphi_n \tag{2.14}$$

In Eq. (2.14), the depth of the sound source is h, the depth of the receiving point is z, and the sound pressure at the receiving point is expressed as the sum of a series of normal modes ϕ_n. Where, the following differential equation is satisfied:

$$\frac{d^2 U_n}{dz^2} + \left(k_1^2 - k_n^2\right) U_n = 0 \quad , \quad 0 < z \le H$$

$$\frac{d^2 U_n}{dz^2} + \left(k_2^2 - k_n^2\right) U_n = 0 \quad , \quad H \le z$$

$$k_1 = \omega/c_1(z), k_2 = \omega/c_2$$

where ω—angular frequency of sound wave. From the boundary conditions, the dispersion equation of the eigenvalue of normal mode wave shows that the phase velocity c_n and wave number of normal wave in horizontal direction depend on the eigenvalues k_n:

$$k_n^2 = \left(\frac{\omega}{c_n}\right)^2 \tag{2.15}$$

So far, various programs have been developed to calculate the normal mode wave field conveniently on a digital computer, so as to estimate the sound intensity and transmission loss in shallow water.

It is well known that there can be any form of vibration on an unbounded string. However, on a bounded string with two fixed ends, the vibration excited by the initial disturbance can only have the form of standing wave, that is, it only contains discontinuous frequency components: fundamental frequency, second harmonic... These frequencies are called normal vibration frequencies, and these fixed vibration forms have specific amplitude distribution shapes along each point of the string, It is called normal vibration. That's why the strings have a particular tone and timbre. Similarly, any form of wave can be excited in the unbounded space, but in the bounded space, that is, the space is limited by the sea surface and seabed, the sound field can only have a specific standing wave form along the depth direction, which is called normal mode wave.

The normal mode wave transmits along the horizontal direction and has the form of standing wave along the depth direction. Figure 2.27 shows the amplitude distribution $U_n(z)$ of normal mode waves in a uniform water layer. The amplitude of sound pressure on the sea surface is zero, and the amplitude distribution of normal mode wave in the water layer is sinusoidal. There is a "tail" which decays exponentially in the seabed medium. The nth normal mode wave has $(n - 1)$ nodes in the water layer. It can be seen from Fig. 2.27 that the energy of normal mode wave is mainly concentrated in the water layer, and only a small part of the energy transmits in the seabed medium. The lower the frequency of sound wave has, the longer the tail of normal wave amplitude distribution curve in the seabed medium is, which means that most of the sound energy of low-frequency sound wave transmits in the seabed. $U_n(h)$ is called the excitation function of normal mode wave, which has the same function form as the amplitude distribution function of normal mode wave. If the sound source and the receiver are placed at the appropriate depth to maximize $U_n(z)$ and $U_n(h)$, the maximum sound pressure of the normal mode wave can be received. If the sound source or receiver is placed at the node of the excitation function or amplitude distribution function of the normal mode wave, the normal mode wave cannot be excited or received. The intensity of normal mode wave excited by sound source also depends on the frequency of sound wave, which is similar to the resonance phenomenon of electrical or mechanical vibration. Figure 2.28 shows the resonance characteristics of normal excitation intensity. It can be seen from the figure that a certain order normal mode wave cannot be excited below a certain frequency, which is called cut-off frequency.

Equation (2.15) can be rewritten as follows:

$$c_n = \omega / k_n \tag{2.16}$$

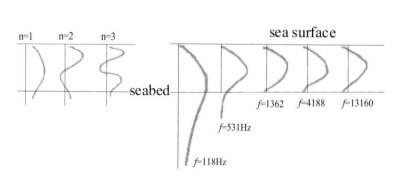

(a) amplitude distribution of normal mode

(b) amplitude distribution of the first normal mode

Fig. 2.27 Amplitude distribution of normal mode waves in uniform water layer

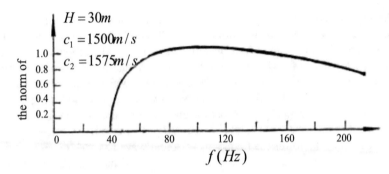

Fig. 2.28 Excitation intensity of the first normal mode wave

where, c_n is the phase velocity of the nth normal mode wave propagating along the direction of r. A normal wave is a cylindrical wave whose isophase surface is a series of coaxial cylinders. c_n represents the propagation velocity of a series of coaxial isophase cylinders, which depends on the root k_n of the dispersion equation. k_n is a complex function of acoustic frequency, so the phase velocity of normal mode wave is also a complex function of frequency. The phase velocity is related to the frequency of sound wave, which is called dispersion phenomenon. For Pekeries model, the typical relationship between phase velocity c_n and frequency f is shown in Fig. 2.29. For the high frequency sound wave, the phase velocity is close to the plane wave phase velocity c_1 in the water layer medium, and for the low frequency sound wave near the cut-off frequency, the phase velocity is close to the plane wave phase velocity in the seabed medium.

Because of the dispersion phenomenon in the shallow water waveguide, if the sound source emits a broadband sound pulse signal, the received signal will be distorted due to the dispersion. Figure 2.30 shows the waveform received after passing through a low-pass filter when the sound source emits an exponential pulse (such as an explosion pulse). The first arrival is the low-frequency sound wave whose frequency is close to the cut-off frequency of the first normal wave. These sound

Fig. 2.29 Phase velocity and group velocity of the first normal mode wave in Pekeries model

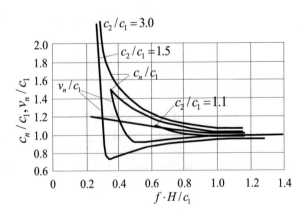

Fig. 2.30 Dispersion waveform of the third normal mode wave

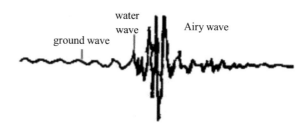

water wave

ground wave

Airy wave

waves have the sound velocity close to the seabed medium, so these low-frequency sound waves are called ground waves. Their excitation intensity is very small, so the amplitude of these low-frequency sound waves is also small. Then comes the water wave, which has a higher frequency and a sound velocity similar to that of sea water. The signal reaching the maximum intensity is called Airy wave, which has medium frequency and minimum velocity.

In order to further explain the dispersion of pulse signal in shallow water, the concept of group velocity should be introduced. The transmission velocity of wave front is called phase velocity, and the transmission velocity of acoustic energy is called group velocity. In colorless dispersion waveguide, phase velocity is independent of acoustic frequency, and the phase velocity is equal to group velocity. In dispersive waveguides, both phase velocity and group velocity are functions of frequency. Generally speaking, phase velocity and group velocity are not equal. An explosion pulse has a wide frequency component, which excites multiple order normal waves. Each order normal wave contains abundant frequency components. Figure 2.30 shows that the sound waves arriving at different times have different frequencies. In other words, the energy transmission velocity (group velocity) of sound waves at different frequencies is different.

Let's consider a certain order normal wave, which is a cylindrical wave, omitting the amplitude factor, and its function is:

$$e^{j(k_n r - \omega t)}$$

The above formula shows that the phase of sound wave is a function of frequency, distance, and time of sound wave, so the phase is constantly changing during the transmission process, and its change amount is $(k_n dr - \omega dt)$. For a pulse signal, if all frequency components near a certain frequency are superposed in phase, the maximum energy state can be obtained

$$dk_n dr - d\omega dt = 0$$

i.e. $$\frac{dr}{dt} = \frac{d\omega}{dk_n} = U_n(w) \qquad (2.17)$$

Fig. 2.31 Theoretical curve
of low frequency
propagation loss in shallow
water. Depth of water $H =$
160 m, $c_1 = 1500$ m/s, $c_2 =$
1550 m/s. $\rho_1 =$
1.2 g/cm^2, seabed absorption $\beta =$
1.0 dB/λ

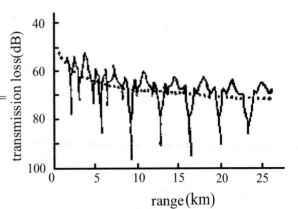

It can be seen from Eq. 2.17 that the group velocity can be calculated only by calculating the eigenvalue. For Pekeries model, the typical results are shown in Fig. 2.29. The results show that the group velocity of low frequency sound wave is close to that of seabed medium, the group velocity of medium frequency sound wave is lower than that of plane wave phase velocity of water medium, and the group velocity of high frequency sound wave is close to that of water medium. These can be used to clearly explain the received waveform shown in Fig. 2.30.

Let's come back to the sound field of simple harmonic point source. The normal mode wave is traveling wave in the direction of r, and the amplitude of each normal mode wave decays according to the law of $1/\sqrt{r}$. When there are multiple normal modes, each normal mode wave will interfere with each other, and the attenuation law of sound pressure with distance will be very complex. An example is shown in Fig. 2.31. We can see a series of interference patterns. The midpoint line in the figure represents the attenuation law of cylindrical wave.

References

1. Tolstoy I, Clay CS. Ocean acoustic. McGraw-Hill Book Company; 1977.
2. Greenspan AM, Tschiegg CE. Sing-around ultrasonic velocimeter for liquids. J.A.S.A. 1959; (31):1038.
3. Knudson VO, Alford RS, Emling JW. Underwater ambient noise. J Marine Res. 1948; (7):410.
4. Urick RJ. Principles of underwater sound, vol. 22, 3rd edn. Los Atlos, California: Peninsula Publishing; 1983. p. 23–4.
5. He ZY, Zhao YF. Theoretical basis of acoustics. Beijing: National Defense Industry Press; 1981.
6. Hui JY, Wang LS. Adaptive matching filter and adaptive correlator. Underwater acoustic communication; 1986.
7. Pekeries CL. Theory of propagation of explosive sound in shallow water. The Geological Society of America; 1948.

Chapter 3
Coherent Multipath Channel

3.1 System Functions of Coherent Multipath Channel

The sonar transmitter or source sends out the sound wave carrying information, and reaches the sonar receiving hydrophone array through the ocean. The sonar system analyzes and processes the received signals, thus determining whether there is a target and what the state parameters are and what target types they have. From the perspective of communication theory, the ocean is the sound channel. It not only transforms the energy of the target signal (acoustic transmission loss), but also transforms the emission waveform of the sound source. Therefore, the acoustic channel can be regarded as a filter for the transformation of the emission waveform. Generally speaking, the acoustic channel is a time-varying and space-varying random channel, so it must be described by time-varying and space-varying random filters. However, in most applications, the experiment shows that the acoustic channel can be regarded as a coherent multi-path channel with slow time-varying. If the observation or processing time is not too long, the acoustic channel can be described by time-invariant filter. Coherent multi-path channel model means: the medium and boundary are time invariant, and the position of sound source and receiver is also determined. The signals from the sound source reach the receiving point in various ways, and they interfere with each other and overlap, thus generating complex spatial interference patterns and complex filtering characteristics. Therefore, the received signal generates distortion and has a significant difference from the emission waveform. Coherent multi path channels are described by time invariant filters.

The acoustic wave is a micro amplitude wave, which satisfies the superposition theorem, so it can be reasonably considered that the coherent multipath channel is linear and can be described by a linear time invariant filter.

This section describes the system functions and characteristics of coherent multipath channels.

Firstly, the basic knowledge of linear network is briefly introduced. The characteristics of linear network can be described by its system function. Given a linear

© Harbin Engineering University Press 2022
J. Hui and X. Sheng, *Underwater Acoustic Channel*,
https://doi.org/10.1007/978-981-19-0774-6_3

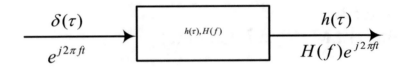

Fig. 3.1 The system function of linear network

four terminal network, two kinds of signals can be used to measure the characteristics of the network. One way is to add a pulse δ to the input. The output waveform of the network is called $h(\tau)$, the impulse response function of the network; The otherway is to add a harmonic signal to the input. The ratio of the output signal to the input signal of the network is called $H(f)$, the transfer function of the network, also known as the frequency response function of the network. The sum of the above two system functions $h(\tau)$ and $H(f)$ is not independent of each other. They are Fourier transform as follows:

$$h(\tau) = \int_{-\infty}^{+\infty} H(f)e^{j2\pi f\tau}\, df$$

$$H(f) = \int_{-\infty}^{+\infty} h(\tau)e^{-j2\pi f\tau}\, d\tau \tag{3.1}$$

The system function of linear network is shown in Fig. 3.1.

Equation (3.1) shows that only one system function can completely determine the characteristics of the network, and only one system function can be obtained from Fourier transform. The two system functions have their own advantages in different applications. $h(\tau)$ can be used to study the waveform transformation of the input signal, while $H(f)$ can be used to study the spectrum transformation of the input signal.

For Eq. (3.1), a brief proof can be stated on the basis of the principle of superposition. Superposition principle is the basic principle of linear system. The superposition principle points out that if the input signal is $z_n(t)$, the corresponding output of the system is $w_n(t)$; When the input waveform is $w_n(t)$, the superposition principle confirms that the output $\sum_{n=1}^{N} z_n(t)$ of the network is:

$$w(t) = \sum_{n=1}^{N} w_n(t) \tag{3.2}$$

According to the properties of Dirac function $\delta(t)$, a signal $z(t)$ with arbitrary waveform can be expressed as:

$$z(t) = \int_{-\infty}^{+\infty} \delta(t - \tau) z(\tau) d\tau = \int_{-\infty}^{+\infty} \delta(\tau) z(t - \tau) \, d\tau \tag{3.3}$$

The physical meaning of Eq. (3.3) is that any signal can be regarded as the sum of a series of δ pulses. The meaning of the first formula is: the appearance time of δ pulse is time $\tau = t$, and its amplitude is $z(t) d\tau$.

For a linear network, when the input signal is $\delta(\tau)$, the output of the network is $h(\tau)$. According to the superposition principle, when the input signal is $z(t)$, the output $w(t)$ of the network is:

$$w(t) = \int z(t - \tau) h(\tau) \, d\tau \tag{3.4}$$

Equation (3.4) omits the infinite integral limit. For the sake of brevity, it is also omitted in the following parts.

If the input signal of the network is harmonic signal $e^{j2\pi ft}$, it can be seen from formula (3.3):

$$e^{j2\pi ft} = \int \delta(t - \tau) e^{j2\pi f\tau} \, d\tau$$

Take $z(t) = e^{j2\pi ft}$ in Eq. (3.4). From the definition of transfer function, we know the output $w(t) = H(f) e^{j2\pi ft}$ of the network at this time, so Eq. (3.4) can be written as:

$$H(f) e^{j2\pi ft} = \int h(\tau) e^{j2\pi f(t-\tau)} d\tau$$

By eliminating the common factor on both sides of the equation, the following result can be obtained:

$$H(f) = \int h(\tau) e^{-j2\pi f\tau} d\tau$$

The above formula is Eq. (3.1). It can be seen that $H(f)$ and $h(\tau)$ are Fourier transforms for each other. After getting Fourier transform on both sides of Eq. (3.4), the following results are obtained:

$$W(f) = Z(f) \cdot H(f) \tag{3.5}$$

where $W(f)$ denotes the output spectrum of network, $Z(f)$ denotes the spectrum of input signal.

Next, we investigate the system functions of multipath coherent channels when the source and receiver are stationary. For the sake of simplicity, the viewpoint of ray acoustics is adopted. The sound signal is emitted from the sound source and arrives at the receiving point along the rays of different paths. The total received signal is the interference superposition of the signals transmitted through all the rays of the receiving point. The signal amplitude arriving along the i path is recorded as A_i. The value of A_i can be calculated by the ray calculation, and the signal delay arriving along the τ_{0i} path can be calculated, ignoring the frequency characteristics of the medium absorption, It is assumed that there is no dispersion in the sound transmission along any path, that is to say, if the sound source emits a δ pulse, the signal arriving along each path will be a delayed δ pulse, and the waveform of the sound signal along each individual path will not change during the transmission process, so the system function of coherent multi-path channel is as follows:

$$h(t) = \sum_{i=1}^{N} A_i \delta(t - \tau_{0i}) \tag{3.6}$$

Equation (3.6) refers to the impulse response function of the channel, that is, the waveform received by the receiving point when the sound source emits a δ pulse. There are N ray paths that have important contributions to the sound field.

Fourier transform is performed on both sides of Eq. (3.6), and we obtain:

$$H(f) = \sum_{i=1}^{N} A_i e^{-j2\pi f \tau_{0i}} \tag{3.7}$$

If the sea depth, the profile shape of the seabed, the sound velocity profile, and the relative geometric relationship between the sound source and the receiving point are known, the concrete form of the coherent multipath channel system function can be obtained from Eqs. (3.6) and (3.7) only by calculating the ray parameters.

An example of the calculation is shown in Fig. 3.2. Under the given conditions, there are 8 channels that have important contributions to the sound field, and the ray

Fig. 3.2 Transfer function example of deep-sea channel. **a** Amplitude frequency characteristics. **b** Phase frequency characteristics

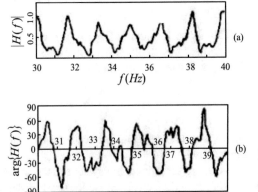

parameters are listed in Table 3.1. This is roughly equivalent to the deep-sea channel at 250n mile.

From this example, we can see that the channel is like a "comb filter", with "passband" and "stop band" appearing alternately, which is called "sub passband". The average width of each sub band is about 1 Hz. The figure also shows that the phase frequency characteristic of the channel transfer function is not linear, which means that the signal waveform will be distorted in the process of transmission. In tourniquet, not only the received signal amplitude is small, but also the signal waveform distortion is serious.

Figure 3.3 shows an example of the transmission function in the shallow channel of 10 km uniform water layer with a sea depth of 100 m. By contrast, the transmission function of the shallow sea coherent multi-channel in the negative gradient aquifer at the same depth and distance is shown in Fig. 3.4. It is found that the average sub band of the channel in the uniform shallow water is wide, and it is about 100–300 Hz in the given case, while for the negative gradient water layer, it is only dozens of Hz wide. Compared with the deep sea, it can be concluded that the thinner the water layer is, the wider the average pass width is.

Because the sound field of the receiver is the result of interference superposition of multi-path arrival, the system function of the channel is very sensitive to the changes of environmental parameters and the relative position of the sound source and receiver. For example, when the thickness of the water layer changes, the relative delay law of the arrival of a series of rays will change accordingly, so the situation of multi-path interference will also change, thus changing the waveform of the received signal and the system function of the channel. Further analysis shows that when the position of the receiver changes dozens of cm in the vertical direction, the system function of the channel changes significantly, while when the receiver moves along the sound transmission direction in the horizontal plane, the system function change of the same order of magnitude needs to move several meters. In other words, the sensitivity order of channel system function to environmental parameters is: vertical

Table 3.1 The ray parameters of deep sea coherent multipath channel

Sound sequence number i	1	2	3	4	5	6	7	8
A_i	0.18	0.14	0.17	0.42	0.65	1.00	0.15	0.29
$\tau_{0i} - \tau_{06}(s)$	−3.0	−2.4	−1.8	−1.2	−0.6	0	0.5	0.9

Fig. 3.3 Example of shallow water transmission function in uniform layer

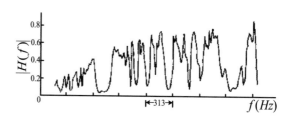

Fig. 3.4 Example of
negative gradient shallow
water transmission function

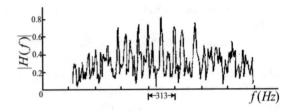

position change, water layer thickness change, horizontal position change, and sound velocity change in water layer.

Because the system function of the channel is very sensitive to the changes of the environmental parameters, it is difficult to predict the details of the system function of the channel accurately, and it has no important engineering application value. However, it is of great significance to study the basic characteristics of the system function of the coherent multipath channel for sonar signal processing.

The basic characteristics of the known system functions of coherent multipath channels are summarized as follows:

1.   The system function of coherent multipath channel is sensitive to the change of environmental parameters, so when the target moves relative to the sonar carrier, the system function of channel is time-varying. In order to adapt to the channel, the sonar signal processor must be time-varying and adaptive.
2.   The transfer function of the channel is like a comb filter, and its phase frequency characteristic is not linear, so the processing effect of the copy correlator in the channel is poor. The width of the average sub band is related to the thickness of the water layer and the sound velocity distribution. The thinner the water layer is, the wider the average sub band is. The uniform water layer and the positive gradient waveguide have wider average sub bands.

3.2 Correlator and Matched Filter

This section discusses the problem of detecting the known signal $s(t)$ under the steady white background interference [1]. Interference is referred to as a stationary process, which means that the statistical characteristics of interference do not change with time. For most practical applications, the stationary process is ergodic, that is, the time average can replace the system average. The so-called white interference means that the power spectrum of interference is constant for all frequencies, and the time correlation radius of white noise is zero, that is, two adjacent time samples are not correlated. The so-called known signal means that the waveform of the signal is completely known except the arrival time and the amplitude of the signal remain unknown. It is theoretically pointed out that the best processor to output the maximum signal to noise ratio under the above conditions is the matched filter. Next, the frequency response function of the matched filter and its output signal to noise ratio are obtained.

Since the interference is assumed to be white, we can set the power spectral density of the input interference to be $N_0/2$, then when the transfer function of the filter is $H(f)$, the output interference power turns out to be:

$$\frac{N_0}{2} \int_{-\infty}^{+\infty} |H(f)|^2 \mathrm{d}f = N_0 \int_0^{\infty} |H(f)|^2 \mathrm{d}f \tag{3.8}$$

According to Eq. (3.5), the output signal power of the filter is:

$$\left| \int_{-\infty}^{+\infty} Z(f)H(f)e^{j2\pi ft}\mathrm{d}f \right|^2 = \left| \int_0^{+\infty} Z(f)H(f)e^{j2\pi ft}\mathrm{d}f \right|^2 \tag{3.9}$$

where, $Z(f)$ is the spectrum of input analytic signal $z(t)$, and its negative spectrum component is zero. Note that the signal to noise ratio of the output power of the filter is d. From Eqs. (3.8) and (3.9), it can be seen that:

$$d = \frac{\left| \int_0^{+\infty} Z(f)H(f)e^{j2\pi ft}\mathrm{d}f \right|^2}{N_0 \int_0^{+\infty} |H(f)|^2 \mathrm{d}f} \tag{3.10}$$

If the transfer function of the matched filter is denoted as $H_0(f)$, then substituting $H_0(f)$ into Eq. (3.10) can make d reach the maximum value, so that solving the system function of the matched filter can be reduced to finding the extreme value of Eq. (3.10). The solution can be obtained by using Schwartz inequation of complex analytic function. Schwartz inequation is as follows:

$$\left| \int_a^b u(x)v(x)\mathrm{d}x \right|^2 \leq \int_a^b |u(x)|^2 \mathrm{d}x \int_a^b |v(x)|^2 \mathrm{d}x \tag{3.11}$$

Only when $u(x) = kv^*(x)$. Basically, the symbol "*" denotes conjugation.

The equation holds, where k is an arbitrary constant. Applying Schwartz inequation to the molecules of Eq. (3.10), we obtain

$$\left| \int_0^{\infty} Z(f)H(f)e^{j2\pi ft_1}\mathrm{d}f \right|^2 \leq \int_0^{\infty} |Z(f)|^2 \mathrm{d}f \int_0^{\infty} |H(f)|^2 \mathrm{d}f \tag{3.12}$$

t_1 is the time when the signal takes the maximum peak power. The first integral on the right side of Eq. (3.12) represents twice the signal energy, and the second integral cancels the denominator in Eq. (3.10), so when

$$H(f) = H_0(f) = kZ^*(f)e^{-j2\pi f t_1}, \, f \geq 0 \tag{3.13}$$

we obtain

$$d = d_{max} = \frac{2E}{N_0} \tag{3.14}$$

where, E is the signal energy. Equation (3.13) shows that the transfer function of the matched filter with the maximum output power signal to noise ratio is the complex conjugate of the signal spectrum, which represents the time delay generated by the filter. The impulse response function of the matched filter can be obtained by Fourier transform of Eq. (3.13).

$$h_0(t) = k \int_{-\infty}^{+\infty} Z^*(f)e^{j2\pi f(t-t_1)} \, df \tag{3.15}$$

i.e.

$$h_0(t) = kz^*(t_1 - t) \tag{3.16}$$

Equation (3.16) shows that the impulse response function of the matched filter is a signal waveform with inverted time axis.

Suppose the bandwidth of input signal and interference of the matched filter is B, the pulse width of the signal is T, and the average power of the signal is S, then the signal to noise ratio gain of the matched filter is:

$$G = \frac{Output \, power \, signal \, to \, noise \, ratio}{Input \, power \, signal \, to \, noise \, ratio} = \frac{2E/N_0}{S/(N_0 \cdot B)}$$

$$= \frac{2BE}{S} = \frac{2BST}{S} = 2BT \tag{3.17}$$

As a result, the output signal to noise ratio gain of the matched filter is twice the product of the effective bandwidth B and the pulse width T.

According to Eqs. (3.4) and (3.16), the output waveform $R_{zz}(t - t')$ of the matched filter signal is:

$$R_{zz}(t - t') = k \int_{-\infty}^{\infty} z(t' - \xi)z^*(t - \xi) \, d\xi$$

Set $\eta = t' - \xi$, and then:

$$R_{zz}(t - t') = k \int_{-\infty}^{+\infty} z(\eta)z^*[\eta + (t - t')] \, d\eta$$

Let $t - t' = \tau$ be the above equation:

$$R_{zz}(\tau) = k \int_{-\infty}^{+\infty} z(t)z^*(t + \tau)\,dt \qquad (3.18)$$

Equation (3.18) is the expression of correlation function of complex signal. When $\tau = 0$, i.e., $t = t'$ the correlation function reaches the peak value. Equation (3.18) shows that the correlator and matched filter are equivalent.

Finally, it is the brief summary of the basic concepts of matched filter:

(1) When the white noise background and the target signal are known, the matched filter is the best linear filter with the maximum output signal to noise ratio, and the output signal to noise ratio is e. When the input background interference is non-white noise, the pre-whitening network can be used to whiten the background interference, and the transfer function of the pre-whitening network is the reciprocal of the interference power spectrum. The signal to noise ratio gain of the matched filter is twice the product of the signal bandwidth and the signal length.

(2) The output power signal to noise ratio of the matched filter is only related to the input signal energy and white noise power spectral density, but not to the fine structure of the signal waveform. The output signal to noise ratio of the matched filter can be increased by enhancing the input signal energy. There are two ways to achieve this goal in sonar technology: one is to increase the sound source level, such as increasing the transmitting power, improving the efficiency of the transmitting transducer, widening the aperture of the transmitting transducer, etc. The second is to increase the signal width. When the sound source level increases to the cavitation limit, cavitation bubbles will be generated in the medium, which will hinder the sound radiation, resulting in the decrease of sound source level and serious distortion of the radiated acoustic signal waveform; The latter approach is often practical and effective, and its limitation is often the size of hardware implementation, for example, the energy storage capacity of transmitter should not be too large, and the processing time of processor cannot be too long due to the limited operation speed of hardware.

Although the output signal to noise ratio of matched filter has nothing to do with the signal waveform (when the signal energy is constant) under ideal conditions. In fact, choosing the appropriate waveform has an important impact on the working performance (detection ability and measurement performance) of sonar. These problems will be discussed in the remaining chapters of this book.

(3) Increasing the signal bandwidth B, the output signal to noise ratio gain of the matched filter will increase, but the output signal to noise ratio will not change (when the signal energy remains constant). However, conscientious readers point out that the theoretical model of matched filter implicitly assumes that signal and interference are additive and independent of each other. Only when BT value is large enough, the above assumption is approximately true. Especially when the interference background is reverberation, it is generated by the

transmitted signal, so generally speaking, the signal and reverberation are not independent of each other. Only when the bandwidth of the transmitted signal is large enough, the signal and reverberation are approximately independent.

(4)   Matched filter and autocorrelator have the same output, so they are equivalent in principle, but the implementation method is different. In particular, it should be noted that the correlator must calculate the whole correlation function rather than just one of its values to ensure the output peak, thus, the correlator must have a time compressor.

With the rapid development of modern digital technology, correlator is easier to implement than matched filter. Correlator is often used in sonar technology.

(5)   The transfer function of the matched filter is the complex conjugate of the signal spectrum, so in the physical sense, the matched filter does two kinds of processing to the signal: one is to remove any nonlinear part of the signal phase frequency function, so that at a certain time all the frequency components in the signal can be superimposed in the same phase at the output end to form a peak value; The other is to filter and weigh the input waveform according to the amplitude frequency characteristics of the signal, so as to receive the signal energy most effectively and suppress the output power of the interference.

3.3   Signal Ambiguity Function

Ambiguity function was originally used to study the measurement and resolution performance of radar. Sonar technology uses this concept. It is not only a basic tool in analyzing the speed and distance of the measurement target, but also very useful in analyzing the detection ability of sonar. This section introduces its basic concepts and will further analyze them in the rest of the book.

Correlator and matched filter are equivalent in basic performance, and their output waveform is the autocorrelation function of input signal, so it is necessary to analyze the basic performance of autocorrelation function. The sonar target is virtually moving, and the echo has Doppler frequency shift, so the correlator needs to calculate the copy of the transmitted signal and the correlation of the echo. In the ideal channel, the waveform of the echo is consistent with the transmitted signal except the Doppler frequency shift.

We assume that the transmitted signal is a narrow-band signal, so it can be represented by a complex analytic signal, denoted as $z(t)$, and its spectrum is $Z(f)$, and then the reference signal of the copy correlator should be $z^*(t + \tau)$ according to the theory in the previous section, where τ represents the search delay; Narrow band signal with Doppler shift can be expressed as $z(t)e^{-j2\pi vt}$, v, with v as Doppler shift, and its spectrum is $Z(f + v)$. Therefore, the principle of copy correlator in ideal channel is shown in Fig. 3.5. Figure 3.6 points out its equivalent matched filter.

We define the autocorrelation function of the signal with Doppler shift and delay difference as the ambiguity function of the signal. This function is the output of

Fig. 3.5 The schematic diagram of copy correlator

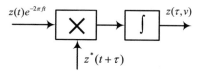

Fig. 3.6 The schematic diagram of matched filter

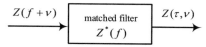

matched filter or copy correlator in ideal channel. The ambiguity function is expressed by $\chi(\tau, v)$, i.e.

$$\chi(\tau, v) \underline{\underline{\Delta}} \int z(t)z^*(t + \tau)e^{-j2\pi vt}\,dt = \int Z^*(f)Z(f + v)e^{-j2\pi f\tau}\,df \qquad (3.19)$$

In Eq. (3.19), the small letters represent the time waveform, the capital letters represent the corresponding frequency spectrum, v represents the Doppler frequency shift generated by the target motion, and τ describes the time delay difference between the target echo and the correlator reference signal. Equation (3.19) is used to describe the detection performance and measurement performance of the copy correlator in the ideal channel. From this formula, we know it is completely determined by the transmitted signal itself. Therefore, one of the limitations of the performance of the sonar system is limited by the selected signal waveform.

It is very intuitive that ambiguity function is used to characterize the measurement and resolution performance of signal or corresponding copy correlator. The following description is helpful for readers to understand this. The peak value of the ambiguity function modulus represents the energy of the signal. We normalize it as one point. The peak value modulus of the normalized ambiguity function is 1. The modulus of Eq. (3.19) is a curved surface. Set the peak value to be at (0,0), that is, the correlator is tuned to $\tau_0 = 0$, $v_0 = 0$), which means that the correlator outputs the maximum signal for the target at a certain distance (τ corresponds to the time delay τ_0) and a certain velocity (corresponding to the fact that Doppler frequency shifts to v_0); For another adjacent target, the distance difference is $\Delta R = c\Delta\tau/2$. c is the sound velocity, and the velocity difference is $\Delta v = \Delta v/0.69 f_0$. The unit of v is Hz, the unit of v is kn, and the unit of f_0 is kHz. Whether the correlator can distinguish the two targets depends on whether the main peak shape of ambiguity function is sharp or not. In other words, the main peak width of the ambiguity function can represent the measurement and resolution performance of the signal, and describes the accuracy limit of the speed and distance of the target measured by the correlator. If the sonar is used to measure the two-dimensional ambiguity function of the target signal in real time, it must adopt the correlator bank (or matched filter bank). Each correlator is tuned to certain Doppler frequency, and a group of correlators cover all possible velocity ranges of the target. In most cases, in order to make the sonar equipment simple, only one correlator is used, which is tuned to zero Doppler. When the target has Doppler and time delay deviation Δv and $\Delta \tau$, the output peak value of the correlator will drop from 1 to $|\chi(\tau_0 + \Delta \tau, v_0 + \Delta v)|$, whose 3 dB decreases to corresponding Δv. It is called the Doppler tolerance of correlator. Δv, the width of the main peak of ambiguity function, not only represents the ability to distinguish moving targets, which is called Doppler resolution, but also represents the ability of single correlator to detect moving targets, which is called Doppler tolerance, as the result of looking at the same thing from different angles.

For narrowband signal modulated by carrier frequency, its measurement and resolution performance depend on the waveform of complex envelope. The modulated narrow-band signal can be expressed as:

$$z(t) = a(t)e^{j\omega_0 t}$$

where, $a(t)$ is the complex envelope and the subcarrier frequency waveform of the signal as well. If the spectrum is $A(f)$, the ambiguity function of the signal can also be written as:

$$\chi(\tau, v) = \int a(t)a^*(t+\tau)e^{-j2\pi vt}dt = \int A^*(f)A(f+v)e^{-j2\pi f\tau}df \quad (3.20)$$

Fuzziness function is a very special kind of function, which will not be proved here, but its important characteristics are listed as follows [1]. It is assumed that the signal is normalized, i.e.

$$\int |s(t)|^2 dt = 1$$

And obtains

$$(1) \quad |\chi(0, 0)| = 1 \quad (3.21)$$

$$(2) \quad |\chi(\tau, v)| \leq |\chi(0, 0)| = 1 \quad (3.22)$$

$$(3) \quad \iint |\chi(\tau, v)|^2 d\tau dv = 1 \quad (3.23)$$

$$(4) \quad |\chi(-\tau, -v)| = |\chi(\tau, v)| \quad (3.24)$$

Features (1) and (2) show that the peak value of ambiguity function appears at the tuning point of correlator; Feature (4) shows that the ambiguity function is symmetric concerning the origin; Feature (3) is the most interesting one, which shows that the volume under the ambiguity function surface is constant. Changing the form of signal

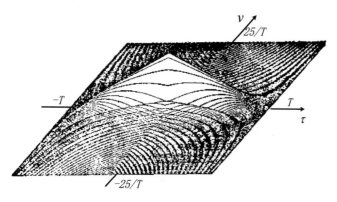

Fig. 3.7 $|\chi(\tau, v)|$ of CW pulse generation

can only change the shape of the ambiguity function surface, but not the volume under the surface.

In order to help readers to get some intuitionistic concepts of ambiguity function, three-dimensional diagrams of ambiguity functions of three kinds of signals are given here. The three signals are as follows:

1. Sinusoid filled rectangular envelope pulse (CW pulse), shown in Fig. 3.7.

$$z(t) = \frac{1}{\sqrt{T}} \text{rect}\left(\frac{t}{T}\right) \cdot e^{j\omega_0 t}$$

$$a(t) = \frac{1}{\sqrt{T}} \text{rect}\left(\frac{t}{T}\right)$$

where, T represents pulse width;
   f_0 represents carrier frequency;

$$\text{rect}\left(\frac{t}{T}\right) = \begin{cases} 1 & |t| \le \frac{T}{2} \\ 0 & others \end{cases}$$

2. LFM signal, shown in Fig. 3.8.

$$z(t) = \frac{1}{\sqrt{T}} \text{rect}\left(\frac{t}{T}\right) e^{j(\pi\mu t^2 + 2\pi f_0 t)}$$

$$a(t) = \frac{1}{\sqrt{T}} \text{rect}\left(\frac{t}{T}\right) e^{j\pi\mu t^2}$$

where, μ represents frequency modulation index.
   The instantaneous frequency of FM signal is:

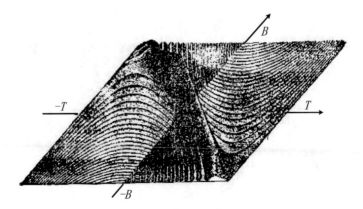

Fig. 3.8 $|\chi(\tau, v)|$ of linear frequency modulation pulse

$$f(t) = \frac{1}{2\pi}\frac{d\varphi(t)}{dt} = \mu t$$

Suppose that the sweep frequency band of instantaneous frequency within the pulse width T is B (frequency modulation bandwidth for short), then:

$$\mu = B/T$$

3.   V-shaped FM signal

$$a(t) = \frac{1}{\sqrt{2}}\left[a_1\left(t + \frac{T}{4}\right) + a_2\left(t - \frac{T}{4}\right)\right]$$

$$a_1(t) = \frac{1}{\sqrt{2}}\text{rect}\left(t/\frac{T}{2}\right)e^{-j\pi\mu t^2}$$

$$a_2(t) = \frac{1}{\sqrt{T/2}}\text{rect}\left(t/\frac{T}{2}\right)e^{j\pi\mu t^2}$$

The prominent bulge of ambiguity function surface is its main peak, and the rest is called skirt. It can be seen from the figure that for CW pulse, its main peak of ambiguity function is relatively flat, its measurement performance and resolution are not ideal, but its skirt is the lowest. The main peak of LFM signal is a thin blade shape, its measurement and resolution are medium, and its skirt is also low. The main peak of V-shaped FM signal is very sharp, its measurement and resolution have good performance, but its skirt is very high with four high "residual ridges". Comparing the above three figures, we can see that the volume under the surface of normalized ambiguity function is constant again. If the main peak is larger, the volume under it is larger, then the skirt is lower; If the main peak is sharp and the volume under it is small, the skirt will bulge significantly and the volume under the skirt appears to be larger (Fig. 3.9).

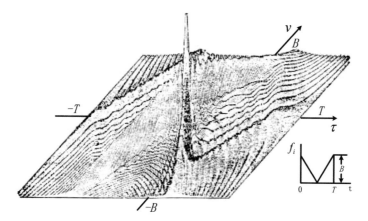

Fig. 3.9 $|\chi(\tau, v)|$ of V-shaped FM pulse

In the ideal channel, it has good measurement and resolution performance when using copy correlator and thumbtack function. The so-called thumbtack function means that the shape of its ambiguity function surface is like the signal waveform of a thumbtack. Its ambiguity function surface has a very sharp main peak, so its skirt is also high. What kind of waveform should sonar choose in the actual ocean acoustic channel? In the early period, it is believed that we should choose the pushpin function. In the mid-1960s, it was proved that due to the complexity of the ocean acoustic channel, especially due to the multi-path interference effect, the thumbtack function was not a suitable sonar waveform for the ocean acoustic channel. If the skirt and residual ridge of thumbtack function are too high, the higher residual ridge will form a series of pseudo peaks through multi-path interference, and their height is even not much different from the main peak. Therefore, the pseudo peaks become significant interference and destroy the obvious signal characteristics, resulting in the decrease of the detection ability of sonar system. On the contrary, it gives up the requirement of measurement and resolution, for example, when the LFM signal is used, its skirt is low, and the multi-path interference will not form obvious pseudo peaks, so its detection performance is better.

3.4 Response of Copy Correlator in Coherent Multipath Channel

The basic time domain processor used in active underwater acoustic equipment is the copy correlator. So far, it has been used in the equipment sonar. The principle of copy correlator is shown in Fig. 3.10. The detected signal is the target echo $s(t)$, which is used as the input signal of correlator together with the superimposed noise; The reference signal is a copy of the transmitting signal. If the transmitting signal is

Fig. 3.10 Schematic
diagram of copy correlator

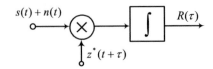

represented as complex analytic signal $z(t)$, the reference signal is $z^*(t + \tau)$. If the transmitting signal does not produce distortion during the channel transmission, the received signal $s(t)$ will be identical to the waveform of the reference signal (copy of the transmitting signal), and $n(t)$ is Gaussian white noise, and the output of the copy correlator has the maximum signal to noise ratio.

At the beginning of 1950s, the application of copy correlator in radar made great achievements, and the range of radar was highly improved, which stimulated foreign underwater acoustic workers to devote themselves to the research of copy correlator in underwater acoustic applications in the next 20 years. The first equipment sonar to use copy correlator was AN/SQS-26, which was tested in the early 1960s. Contrary to the designer's desire, the complex and exquisite two-dimensional ambiguity function real-time display and correlator group did not bring magical detection effect. This sonar adopts CW pulse, LFM signal and pseudo-random coded pulse; The last two kinds of signals are thumbtack signals. Their ambiguity function has sharp main peak, a series of disordered side lobes and high skirt. The sea test shows that the detection effect of the complex thumbtack function and its correlator is not even as good as that of CW pulse and energy detector (square detector integrator after narrowband filtering), and only the linear frequency modulation signal and its correlator have mutual advantages with CW pulse energy detector.

The reason why the detection effect of copy correlator is not ideal has been open to public today. Underwater acoustic channel is a more complex multi-path coherent channel than radar channel. When the target is moving, the channel is still time-varying, and the total received signal waveform is significantly different from the transmitted signal, Therefore, the correlation coefficient between it and the reference signal (the copy of the transmitted signal) is low, which leads to the significant decrease of the gain of the correlator, and the side lobe of the output of the correlator is significantly increased and disordered due to the multipath effect, even the recognition and detection features of the signal are lost. In this section, the response of copy correlator in multipath coherent channel is analyzed.

Figure 3.11 shows how the copy correlator works in the channel. The channel in the figure is regarded as a filter, which should be a two-way channel including the

Fig. 3.11 The working
principle of copy correlator
in time invariant channel

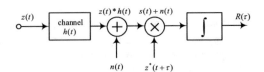

target channel. Without losing generality, for convenience, the channel can still be treated as one-way, and its impulse response function $h(t)$ can be seen from Eq. (3.6) as follows:

$$h(t) = \sum_{i=1}^{N} A_i \delta(t - \tau_{0i})$$

The target echo obtained by the receiving hydrophone should be:

$$s(t) = z(t) * h(t) = z(t) * \sum_{i=1}^{N} A_i \delta(t - \tau_{0i})$$

$$= \sum_{i=1}^{N} A_i z(t - \tau_{0i}) \tag{3.25}$$

where "$*$" represents convolution operation. The output signal components of the copy correlator should be:

$$R(\tau) = \langle s(t) \cdot z^*(t + \tau) \rangle \tag{3.26}$$

where, $\langle \cdot \rangle$ describes the average of the system, and Eq. (3.26) is the signal component output by the correlator. Substituting Eq. (3.25) into Eq. (3.26), we get the following result:

$$R(\tau) = \sum_{i=1}^{N} A_i \langle z(t - \tau_{0i}) z^*(t + \tau) \rangle = \sum_{i=1}^{N} A_i \chi(\tau + \tau_{0i}, 0) \tag{3.27}$$

It can be seen from Eq. (3.27) that in a coherent multipath channel, if the main peak of the ambiguity function of the transmitted signal is so sharp (such as the thumbtack function) that the time delay difference of different channels can be distinguished, the correlator cannot make full use of the total energy of the arrival of the multipath signal. The output of the copy correlator is multimodal, and its amplitude depends on the energy of the arrival of a single channel. In addition, the response of high skirt and side lobe or residual ridge to multi-path signal will result in a chaotic interference pattern. Therefore, for the sonar system with copy correlator, the waveform with low ambiguity function and moderate delay resolution should be selected. The bandwidth of the signal should not be too wide. The LFM signal is a suitable waveform.

3.5　Adaptive Correlator

The research of time domain coherent processor is the basic subject of active sonar signal processing. In order to improve the deficiency of copy correlator, a new kind of time domain coherent processor is proposed, which is called adaptive matched filter and adaptive correlator [2].

Since the ocean acoustic channel is complex, its system function is very sensitive to the environmental parameters. Since the target is moving, and the channel is time-varying and space-varying, the processor must be adaptive, in order to make the sonar system match the channel. Only by extracting the channel information in real time and adjusting the signal processor adaptively can the detection effect be good.

How to improve the detection effect of copy correlator? If a reference signal can be generated adaptively in real time, and its waveform is consistent with the detected signal $s(t)$, and only a small interference component is superimposed, the detection effect of the correlator will be fundamentally improved. It seems to be difficult, but the adaptive filter provides this possibility.

The principle block diagram of adaptive correlator is shown in Fig. 3.12. Considering the acoustic channel as a network, its characteristics can be described by its impulse response function $h(\tau)$ or transfer function $H(f)$.

If the transmitted signal is $z(t)$, the received signal is:

$$s(t) = z(\tau) * h(\tau)$$

The spectrum of the received signal is as follows:

$$S(f) = Z(f) \cdot H(f)$$

Make the copy $z(t)$ of the transmitted signal pass through a filter $h_0(\tau)$. If the system function of the filter is exactly the same as that of the acoustic channel, the reference signal $x(t)$ of the correlator must be exactly the same as the target signal $s(t)$. That is, when

$$\left. \begin{array}{l} h_0(\tau) = h(\tau) \\ H_0(f) = H(f) \end{array} \right\} \tag{3.28}$$

the correlator in Fig. 3.12 has the maximum output signal to noise ratio.

Fig. 3.12 Schematic diagram of adaptive correlator

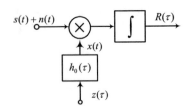

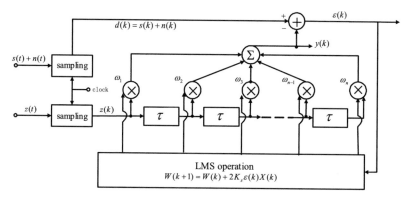

Fig. 3.13 The schematic diagram of adaptive canceller

Filter h_0 is certainly not a conventional filter, it should be an adaptive filter.

Figure 3.13 is the schematic diagram of an adaptive canceller, which uses discrete signal operation. When the reference signal $z(t)$ is a copy of the transmitted signal, the adaptive canceller constitutes an adaptive matched filter, $y(k)$ is the output of the matched filter, and $y(k)$ can adaptively keep consistent with $s(k)$ in the sense of minimum mean square error. $y(k)$ can be used as the reference signal $x(t)$ of the adaptive correlator. The characteristics of the adaptive filter constructed by LMS operation can be adaptively consistent with the network characteristics of the channel. In order to illustrate this point, this paper briefly introduces the basic principle of adaptive canceller.

$z(k)$ is called the input reference signal of the canceller, and $d(k)$ is called the required response,

$$d(k) = s(k) + n(k)$$

$y(k)$ is called the output signal of the canceller, and $\varepsilon(k)$ is called the error between the required response and the output signal.

$$\varepsilon(k) = d(k) - y(k) = s(k) + n(k) - y(k)$$

τ is the unit delay section, $\omega_i (i = 1, 2, \cdots n)$ is the adjustable weight of the adaptive filter, which is adjusted by LMS operation. LMS operation is essentially a steepest descent method. The adjustment direction of weight goes along the estimation gradient direction of error, which will make $\varepsilon^2(k)$ tend to the minimum.

The basic formula of LMS operation is as follows:

$$W(k + 1) = W(k) + 2K_s \varepsilon(k) X(k) \tag{3.29}$$

The weight vector $W(k)$ is as follows

$$W(k) = [\omega_1, \omega_2, \cdots, \omega_n]^T$$

As can be seen from Fig. 3.13:

$$y(k) = \sum_{i=1}^{n} \omega_i(k)z(k - i + 1) \tag{3.30}$$

Equation (3.30) is actually a discrete convolution form, so $y(k)$ can be regarded as the output of the adaptive filter when the input signal is $z(k)$. The impulse response of the filter is determined by the weight vector $W(k)$. K_s is a constant, which is called the "learning" step. If the K_s value is properly selected, the adaptive canceller will converge and adjust the weight vector according to Eq. (3.29) to minimize the mean square error $\overline{\varepsilon^2(k)}$. When the "learning" process is over, that is, when the offset is stable, $y(k)$ will be consistent with $s(k)$ in the sense of minimum mean square error, and then it agrees well with the Wiener Hoff equation:

$$\mathbf{R}_{zz}(k)\mathbf{W}_{LMS}(k) = \mathbf{R}_{zd}(k) \tag{3.31}$$

where $\mathbf{R}_{zz}(k) = \langle \mathbf{Z}(k)\mathbf{Z}^T(k) \rangle$ is the reference signal correlation matrix with $n \times n$ dimension, $\mathbf{R}_{zd}(k) = \langle \mathbf{Z}(k)d(k) \rangle$ is the cross-correlation column vector of response and reference signal, with $n \times 1$ dimension, $\mathbf{Z}(k) = [z(k), z(k - 1), \cdots, z(k - n)]^T$

By making z-transformation on Eq. (3.31), it can obtain:

$$L_{zz}(z)H_0(z) = L_{zd}(z) \tag{3.32}$$

We use $j\omega$ to replace z in Eq. (3.32), and obtain the transfer function $H_0(f)$ in which $L_{zz}(f)$ is the power spectrum of the reference signal, $L_{zd}(f)$ is the cross power spectrum.

From Eq. (3.32), we obtain:

$$H_0(f) = \frac{L_{zd}(f)}{L_{zz}(f)} \tag{3.33}$$

When assuming that $z(t)$ and $n(t)$ are independent of each other, we obtain:

$$H_0(f) = \frac{L_{zd}(f)}{L_{zz}(f)} = \frac{L_{zs}(f)}{L_{zz}(f)} = \frac{L_{zz}(f)H(f)}{L_{zz}(f)} = H(f) \tag{3.34}$$

So far, we have proved that Eq. (3.28) holds. Also, it is proved the characteristics of the adaptive filter are adaptively consistent with the network characteristics of the channel, and the adaptive matched filter extracts the information of the network characteristics of the channel in real time.

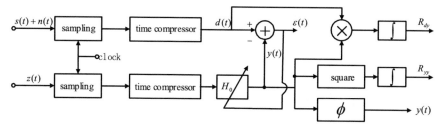

Fig. 3.14 The principle diagram of adaptive copy correlator

The adaptive correlator is practical in engineering implementation. Its principle block diagram is shown in Fig. 3.14. After amplification, filtering, normalization and beamforming, the received signal is transformed into discrete signal by sampling switch. Because the arrival time of the signal is unknown, like any correlator, the adaptive correlator must have a time compressor. The output $y(t)$ is the output of adaptive matched filter and R_{dy} is the output of adaptive correlator.

3.6 Response of Adaptive Correlator in Coherent Multipath Channel

The reference signal of the adaptive correlator will be consistent with the waveform of the input target signal in the sense of minimum mean square error, especially when the number of weights n is large enough and there is no interference. The reference signal will be the same as the input target signal, so the adaptive correlator can effectively match the channel and make full use of all the energy of multi-path arrival.

According to Eq. (3.25), the received target signal is as follows:

$$s(t) = z(t) * h(t) = \sum_{i=1}^{N} A_i z(t - \tau_{oi})$$

Accordingly, the output signal components of the adaptive correlator are as follows:

$$R_{ss}(\tau) = \sum_{i=1}^{N} \sum_{j=1}^{N} A_i A_j \langle z(t + \tau - \tau_{oi}) z^*(t - \tau_{oj}) \rangle$$

$$= \sum_{i=1}^{N} \sum_{j=1}^{N} A_i A_j \chi(t + \tau_{oj} - \tau_{oi}) \tag{3.35}$$

Equation (3.35) can be rewritten as follows:

$$R_{ss}(\tau) = \sum_{j=1}^{N} A_i^2 \chi(\tau) + \sum_{i=1}^{N} \sum_{j=1(i \neq j)}^{N} A_i A_j \chi(\tau + \tau_{oj} - \tau_{oi}) \qquad (3.36)$$

The first term on the right of Eq. (3.36) is the principal component, and the second term is a small component. When $\tau = 0$, the first term is the sum of the energy of the signals from all the channels, which indicates that the adaptive correlator uses the total energy of all the channels, so its detection effect is better than that of the copy correlator.

The problem of active sonar detection is a two-way selective detection, that is, sonar should distinguish one of two situations: one is, there is interference of target signal and background at the input end; The other is, there is only interference at the input. Different from the copy correlator, the reference signal amplitude of the adaptive correlator is very different in the above two cases. When the transmitted signal and interference are independent of each other, and when there is only interference at the input, the output $y(t)$ of the ideal adaptive canceller will be zero, and the output of the adaptive correlator will be zero as well. Of course, the cross-correlation between the transmitted signal and the interference will not be truly zero, and the canceller has various errors. The integral time length of the adaptive correlator is limited. Because of the above reasons, even if the two are statistically independent, $y(t)$ will not be zero, just small. Because the reference signal of the adaptive correlator is smaller when there is only interference at the input, the adaptive correlator has stronger interference suppression ability than the copy correlator.

In the summer of 1984, a one-way reservoir experiment of adaptive correlator was carried out. The water depth is 12 m, the ship is anchored and the transducer is suspended in the water. The center frequency of the transmitted signal is 15 kHz, the pulse width is 200 ms, and the frequency modulation bandwidth is 200 Hz.

The received signal is filtered and recorded on a magnetic tape.

The signal record is replayed in the laboratory. The subcarrier waveform of the signal is taken out after mixer. After being filtered by a low-pass filter with the same bandwidth as the signal, the adaptive correlation analysis and copy correlation processing are carried out by computer.

Table 3.2 shows the comparison of normalized correlation coefficients without interference. It can be seen from the table that the output signal of the adaptive

Table 3.2 Comparison of correlation peaks between adaptive correlator and copy correlator without interference

Signal sequence number		1	2	3	4	5	6	7	8	9	10
Bandwidth 200 (Hz)	Adaptive correlator copy correlator	0.63	0.63	0.44	0.59	0.63	0.51	0.60	0.63	0.62	0.46
		0.19	0.21	0.14	0.20	0.18	0.12	0.11	0.25	0.20	0.27
Bandwidth 400 (Hz)	Adaptive correlator copy correlator	0.55	0.70	0.69	0.62	0.52	0.59	0.57	0.44	0.63	0.72
		0.35	0.28	0.51	0.30	0.45	0.22	0.25	0.39	0.34	0.51

correlator is 3~10 dB higher than that of the copy correlator. Experiments show that the adaptive correlator makes full use of the energy of multi-path arrival.

Figure 3.15 shows the output waveform of the adaptive matched filter in the presence of interference. It can be seen that for the main part of the signal, the output waveform of the adaptive matched filter is consistent with the received signal. The output waveform of the adaptive matched filter has a rising front, which indicates that the learning time of the adaptive matched filter is about 30 ms, which is quite satisfactory.

Figure 3.16 compares the output waveforms of the copy correlator and the adaptive correlator. It can be seen that under the experimental conditions, the copy correlator cannot give obvious signal characteristics, while the output of the adaptive correlator still has obvious correlation peaks.

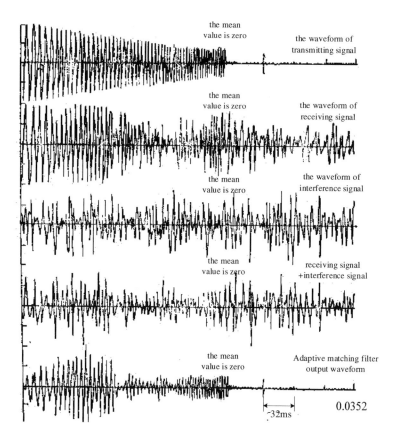

Fig. 3.15 The diagram of waveform comparison (The parameters of transmitting signal: pulse width is 200 ms, LFM bandwidth is 400 Hz)

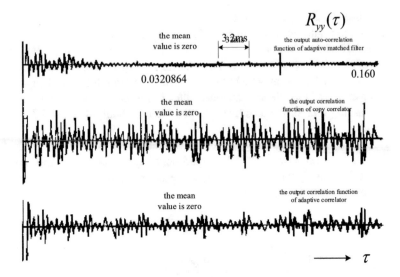

Fig. 3.16 The waveform comparison of correlation function

The data listed in Table 3.3 show that the detection performance of adaptive correlator and adaptive matched filter is significantly better than that of copy correlator, and the signal to noise ratio gain even exceeds the theoretical limit of copy correlator. When the input signal to noise ratio decreases, the gain of the adaptive correlator decreases, but it is still much higher than that of the copy correlator.

3.7 Cross Correlation in Coherent Multipath Channel

The signal processor of sonar system is essentially a spatiotemporal joint processor, and the cross correlator widely used in passive sonar is a typical example. The principle of cross correlator is shown in Fig. 3.17. The output of two hydrophones which are separated in space is amplified, filtered and delayed appropriately, then they multiplied and integrated to get the cross-correlation output.

When the two receivers are separated sufficiently, the local interference is usually uncorrelated, and the correlation number of the sound signal from the source is large. The cross correlator uses the difference of the cross-correlation coefficient to detect the interested target. The application of cross correlator in sonar technology achieves great success, so it is necessary to discuss the cross-correlation of sound field in channel. In particular, it is very important to study the cross-correlation of sound field in the channel not only for the analysis of cross correlator, but also for sonar technology.

Table 3.3 Comparison of detection performance between copy correlator and adaptive correlator (reservoir experimental results)

Signal sequence number	1		2		3		4		5		6		7		8		9	
Input signal to clutter ratio (dB)	3.5	−9.5	2.8	−3.2	−2.0	−6.0	0.29	−1.44	10.5	−7.5	11.4	−1.5	4.1	−2.0	−6.0	−12.0	−2.2	−5.9
Output SNR(dB) — Adaptive matching	20.6	5.1	18.9	/	12.3	7.9	13.2	12.6	35.2	4.1	30.9	10.3	24.1	12.3	7.9	3.2	15.7	6.5
copy correlation	2.1	/	/		1.7	1.9	0.13	0.76	11.3	/	13.5	0.75	8.3	1.7	1.2	/	3.3	/
adaptive correlation	19.2	4.5	17.3	7.4	13.0	9.2	13.8	12.3	37.4	4.6	32.5	11.3	15.1	13.0	7.3	2.5	15.2	7.8
SNR gain (dB) — Adaptive matching	17.1	14.6	15.5	12.7	14.2	13.8	12.9	14.0	24.4	11.7	19.5	11.7	20.0	14.2	13.8	15.2	17.9	12.4
copy correlation	/	/	/	/	3.6	/	/	/	0.7	/	1.9	/	4.23	3.2	/	/	5.6	/
adaptive correlation	15.7	14.2	14.5	10.6	15.0	15.1	13.5	13.7	26.9	12.1	21.1	12.7	21.1	14.9	13.7	14.3	17.4	13.7
Note	Bandwidth 200 (Hz)				Bandwidth 400 (Hz)													

Fig. 3.17 The schematic
diagram of cross correlator

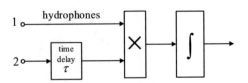

We assume that there is a point source in the coherent multipath channel, and the acoustic signal radiated by it is $z(t)$. When we investigate two receiving points at a long distance, their vector diameters are $\mathbf{r}$ and $\mathbf{r} + \boldsymbol{\rho}$ respectively, and their received signals are $s_1(\mathbf{r}, t)$ and $s_2(\mathbf{r} + \boldsymbol{\rho}, t)$ respectively. When the signal is expressed by a complex function, the spatiotemporal cross-correlation coefficient can be defined as:

$$R_{12}(\boldsymbol{\rho}, t) = \langle s_1(\mathbf{r}, t)s_2^*(\mathbf{r} + \boldsymbol{\rho}, t + \tau)\rangle$$
$$= \overline{s_1(\mathbf{r}, t)s_2^*(\mathbf{r} + \boldsymbol{\rho}, t + \tau)} \tag{3.37}$$

The bar above the character in Eq. (3.37) indicates the average time. Assuming that the process satisfies the ergodic law of each state, the ensemble average and the time average are equivalent.

The theoretical model of spatiotemporal correlation of point source sound field is shown in Fig. 3.18.

From the point of view of acoustic channel theory, the ocean between sound source and receiver 1 is regarded as a filter, and its system functions are $h(\mathbf{r}, t)$ and $H(\mathbf{r}, f)$. The ocean channel between sound source and receiver 2 is regarded as another filter $h(\mathbf{r} + \boldsymbol{\rho}, t)$ and $H(\mathbf{r} + \boldsymbol{\rho}, f)$. The cross-correlation between two points in coherent channel is the cross-correlation of the output of two filters. Therefore, there are:

$$s_1(\mathbf{r}, t) = z(t) * h(\mathbf{r}, t) = \int Z(f)H(\mathbf{r}, f)e^{j2\pi f t}\, df \tag{3.38}$$

$$s_2(\mathbf{r} + \boldsymbol{\rho}, t) = z(t) * h(\mathbf{r} + \boldsymbol{\rho}, t) = \int Z(f)H(\mathbf{r} + \boldsymbol{\rho}, f)e^{j2\pi f t}\, df \tag{3.39}$$

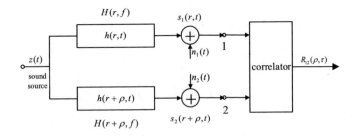

Fig. 3.18 Coherent channel cross correlation model

The cross correlation function is as follows:

$$R_{12}(\boldsymbol{\rho}, t) = \langle s_1(\mathbf{r}, t)s_2^*(\mathbf{r} + \boldsymbol{\rho}, t + \tau)\rangle \tag{3.40}$$

Substituting Eqs. (3.38) and (3.39) into Eq. (3.40) is as follows:

$$
\begin{aligned}
R_{12}(\boldsymbol{\rho}, t) &= \langle z(t) * h(\mathbf{r}, t) \cdot z^*(t + \tau) * h^*(\mathbf{r} + \boldsymbol{\rho}, t + \tau)\rangle \\
&= \langle z(t) \cdot z^*(t + \tau)\rangle * \langle h(\mathbf{r}, t)h^*(\mathbf{r} + \boldsymbol{\rho}, t + \tau)\rangle
\end{aligned}
\tag{3.41}
$$

The right side of Eq. (3.41) can be expressed as a Fourier expression. It is noted that the correlation function and power spectrum are Fourier transforms. Equation (3.41) can be expressed as:

$$R_{12}(\boldsymbol{\rho}, \tau) = \int \langle Z(f)Z^*(f)\rangle H(\mathbf{r}, f)H^*(\mathbf{r} + \boldsymbol{\rho}, f)e^{j2\pi f\tau}\, df \tag{3.42}$$

In a coherent multipath channel, the impulse response function of the channel can be obtained from Eq. (3.6):

$$h(\mathbf{r}, t) = \sum_{i=1}^{N} A_i\delta(t - \tau_{oi})$$

$$h(\mathbf{r} + \boldsymbol{\rho}, t) = \sum_{j=1}^{N} A_j'\delta(t - \tau_{0j}')$$

Substituting the above two formulas into Eq. (3.40), the following formula can be obtained:

$$
\begin{aligned}
R_{12}(\boldsymbol{\rho}, t) &= \sum_{i=1}^{N}\sum_{j=1}^{N} A_i A_j' \langle z(t - \tau_{0i})z^*(t + \tau - \tau_{0j}')\rangle \\
&= \sum_{i=1}^{N}\sum_{j=1}^{N} A_i A_j' \overline{z(t - \tau_{0i})z^*(t + \tau - \tau_{0j}')} \\
&= \sum_{i=1}^{N}\sum_{j=1}^{N} A_i A_j' \chi(\tau + \tau_{0i} - \tau_{0j}')
\end{aligned}
\tag{3.43}
$$

where, $\chi(\tau + \tau_{0i} - \tau_{0j}')$ is the correlation function of sound source radiation signal, or the ambiguity function of zero Doppler frequency offset. The main peak width of the autocorrelation function or the time delay resolution of the ambiguity function are approximately inversely proportional to the bandwidth of the sound source radiation signal. Therefore, Eq. (3.43) shows that the cross-correlation in the channel is related not only to the transfer function of the acoustic channel, but also to the signal radiated by the source, especially to the bandwidth of the latter. We emphasize that when we talk about the cross-correlation of sound field, we must pay attention to the specific conditions concerned, such as signal bandwidth, integration time length and marine environmental conditions.

In the coherent multipath channel, the sound pressure waveforms of any two points are totally coherent, i.e., there is a deterministic transformation relationship between them, but they can be of little correlation. The correlation number is the characterization of the similarity degree of two waveforms. Only when two identical waveforms have their normalized cross-correlation coefficient, it can be equal to 1. For two separate receivers, when receiving the point source radiation signal with limited bandwidth, different waveforms can be received at different receivers due to different multipath interference. Therefore, when the two receivers are separated gradually, the normalized cross-correlation coefficient will decrease. Generally, when the normalized correlation number is reduced to 0.5, the distance between two receiving points is called spatial correlation radius. The horizontal correlation radius perpendicular to the sound propagation direction is called the transverse correlation radius, the horizontal correlation radius along the sound transmission direction is called the longitudinal correlation radius, and the vertical correlation radius is called the vertical correlation radius. The vertical correlation radius is the smallest. The reason is that the vertical direction of the sound field is standing wave, and the change scale of the spatial pattern along the depth direction is far less than the wavelength, so the vertical correlation radius is far less than the wavelength; In the fully coherent open channel, the sound field is basically cylindrically symmetric, so the two receiving points separated horizontally symmetrically perpendicular to the sound propagation direction should receive the same two waveforms, and the transverse correlation radius should be infinite. In the actual ocean channel, the transverse correlation radius is not infinite, but very large, because the actual ocean environment always has a variety of random factors, and various random processes of sound transmission limit the transverse correlation radius. Ocean experiments show that the transverse correlation radius of single frequency wave is indeed very large, which provides sound evidence that the actual marine channel can be regarded as coherent multipath channel under certain conditions. In Chap. 5, more experimental evidence will be provided to support the above view, that is, under the condition of the same width, when the fact that sound transmission moves forward is concerned, the actual ocean acoustic channel can be regarded as coherent multipath channel.

Table 3.4 gives an example of the experimental results of the transverse cross-correlation coefficient of the shallow water sound field. When performing the experiment, the sound source and the receiver are fixed on the seabed, the distance is several km, and the two hydrophones are separated by 47 m. The data in Table 3.4 show that the normalized cross-correlation coefficient is greater than 0.9 as long as the integration time is not too long. It can be seen that the channel is stable enough in at least 20 s, and the spatial correlation radius is at least greater than 47 m. When the integration time is more than 1200 s, the correlation number is significantly smaller, that is to say, the randomness of the marine environment cannot be ignored when the integration time is very long. In fact, the ocean acoustic channel should be regarded

Table 3.4 The experimental results of spatial correlation number of single frequency sound waves in shallow water

Numbers	1	2	3	4	5	6	7	8	9	10	11	12	13	14	15	16
Frequency (kHz)	3	3	3	3	3	3	3	3	1.5	1.5	1.5	1.5	1.0	1.0	1.0	1.0
Correlation integration time (s)	20	20	1200	1200	20	20	1200	1200	20	20	2400	2400	20	20	2400	2400
Cross-correlation coefficients	0.96	0.95	0.32	0.17	0.95	0.94	0.30	0.20	0.84	0.78	0.59	0.50	0.90	0.90	0.35	0.25

Annotation This information is extracted from the proceedings of NATO 1976 advanced Symposium on signal processing

Fig. 3.19 The spatial correlation coefficients (shallow water)

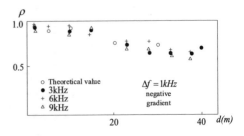

as a slow time-varying coherent multipath channel, and can be regarded as a coherent multipath channel in a very wide range of conditions.

Figure 3.19 shows the marine measurement results of spatial transverse cross-correlation coefficient of narrow-band noise field with 1 kHz bandwidth. The integral time constant is only a few ms. The results show that when the distance between hydrophones is more than 20 m, andthe correlation number is still quite large.

Because the ocean acoustic channel is influenced by many random processes objectively, different integration time constants count the influence of different random processes, so the measured correlation numbers are different. When the integral time constant is short, the channel can always be regarded as coherent. In the coherent multipath channel, the correlation number of two receivers is less than 1, which is not due to the influence of random factors, but the different circumstances of multipath interference, so the signal waveform is different. At this time, the larger the signal bandwidth is, the smaller the spatial correlation radius will be.

The theoretical and experimental results of vertical cross-correlation function of low-frequency sound field in shallow water are shown in Fig. 3.20. The center frequency of the sound wave is 147.8 Hz, the signal bandwidth is 10 Hz, the distance between the sound source and the receiver is 5.5 km, and the water depth is 22.6 m. The consistency of experiment and theory is acceptable. The results show that the vertical correlation radius is only 6~7 m, which is significantly smaller than the wavelength. When the water depth increases, the vertical spatial correlation radius will decrease, and the ordinate in the figure is the normalized polarity coincidence correlation coefficient.

American scholars Tolstoy and clay have calculated the longitudinal cross-correlation function of coherent multipath channels with wave theory. Their results

Fig. 3.20 The vertical correlation coefficients of shallow water sound field Δ—polarity coincidence correlation coefficient (experimental results)

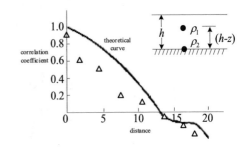

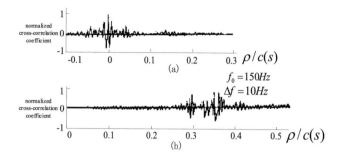

Fig. 3.21 The longitudinal cross correlation function in shallow water coherent multipath channel (a) $\tau = 0$; (b) $\tau = 0.3$, ρ/c—Distance between receiving points; c—Sound velocity。

are shown in Fig. 3.21 [3]. The parameters used in the calculation are: the water depth is 22.6 m, the working frequency is 150 Hz, and the signal bandwidth is 10 Hz. Figure 3.21a is equivalent to the case of $\tau = 0$; Figure 3.21b is equivalent to the case of $\tau = 0.3s$. There are three order normal modes which contribute to the sound field. In the case of $\tau = 0$, the shape of the correlation function is sharp. In the case of $\tau = 0.3s$, the shape of the correlation function has three main bumps, and its maximum peak is not at $\rho/c = 0.3s$, because the phase velocity and group velocity of the three normal modes are different. The dispersion phenomenon complicates the shape of the correlation function, and makes errors when using the correlator to measure the delay. Their theoretical analysis also shows that the longitudinal correlation radius is usually very large.

3.8 Normal Wave Vertical Filtering for Near Field Interference Suppression [4]

Without distinction, we cannot recognize things. Only by making full use of the difference between interference and signal can we detect signals in interference background most effectively, which include the differences in space, time domain, frequency domain and its joint domain.

In the classical sonar equation, the spatial gain of array in detecting point source signal in isotropic noise field is described by the directivity function and aggregation coefficient of array. This concept is correct in unbounded free sound field, so the spatial gain of array has difference between interference and target signal in incident direction. In the coherent channel with boundary, the concept of horizontal directivity of array can be used, but the concept of vertical directivity is not applicable and should be modified.

The horizontal directivity of the array can be used to distinguish the target sound source from the interference source in different directions. Even if the interference source and the target source are in the same direction, and their power spectra are not different, there are still differences in distance and depth between the interference source and the target signal source. This section will show that these differences can also be used to improve the detection performance of sonar.

Clearly speaking, suppose that the interference source and the target sound source are both point sources, and there is no difference in the spectrum or power spectrum. Their position can be the same, and the channel is coherent. A vertical linear array is used to detect the target. Why can vertical linear array suppress short range interference? How can interference be suppressed? What is the processing gain?

There is a distance difference between the target sound source and the interference source. For the point source, all kinds of normal modes contribute to the sound field, or from the point of view of the ray theory, all kinds of rays contribute to the sound field, so the sound field of the point source has a complex standing wave form in the depth direction; For the far-field of point source, only the first or the first normal modes contribute to the sound field, so the standing wave form of the far-field of point source in the depth direction is relatively simple. Therefore, when the receiving sensitivity distribution function is the same as the standing wave function in the depth direction of the sound field of the target signal, the remote target signal can be effectively received and the near-field interference can be suppressed. In fact, it uses spatial filtering and matching technology.

It is more acceptable for engineers to describe the above basic principles from the perspective of acoustic channel theory, seen in Fig. 3.22. The ocean between the sound source and the vertical linear array is regarded as a filter. They are far-field spatial filter and near-field spatial filter. Their transfer functions are $H(r_s, z, \tau$ and $H(r_n, z, \tau)$, r_s and r_n is the distance from the target sound source and interference point source to the receiving point, and z is the depth coordinate of each receiving unit of the vertical linear array. The output of the two filters is superimposed by the sound field, and is represented by an adder in the figure. The output of the adder is $S(f)H(r_s, z, f)$ and $N(f)H(r_n, z, f)$. We mainly consider the problem of spatial filter matching. We can assume that $S(f)$ and $N(f)$ is the same. It is easy to know from the matched filter theory that in order to match the signal, the vertical linear array can be regarded as a spatial filter, and the maximum signal energy can be output by

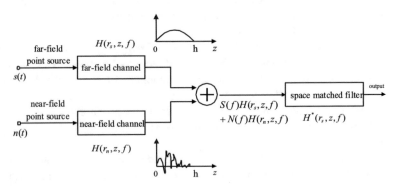

Fig. 3.22 The principle of space matched filter

Fig. 3.23 Near field
interference source and far
field signal source in shallow
water channel

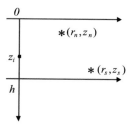

taking the sensitivity distribution function of the vertical linear array as $H^*(r_s, z, \tau)$.
For the far-field first-order normal wave, $H(r_s, z, \tau)$ is a relatively simple function
with prior knowledge. Its functional form is shown in Fig. 2.27, so it is possible to
be realized in engineering.

Next, we will make a brief mathematical description of the above principle of
extracting spatial gain and predict the size of spatial gain. The theory may be unfa-
miliar and boring to engineers, but readers can ignore some details and focus on
understanding the theoretical concepts of sound field and spatial matching. The
following theory [4] is proposed by Shang Erchang, a professor from Institute of
acoustics, Chinese Academy of Sciences.

In the shallow water channel shown in Fig. 3.23, the interference point source and
signal source are located at (r_n, z_n) and (r_s, z_s), the vertical linear array is located at
$r = 0$, and the coordinates of the array elements are recorded as z_i.

We define the spatial gain G of the matrix as:

$$G = (S/N)_L / \overline{(S/N)_0} \tag{3.44}$$

where, $(S/N)_L$, $(S/N)_0$ are the signal to noise ratio of the vertical linear array and
the point receiver respectively, and the horizontal bar on the character represents the
average along the depth. Here we have:

$$\left(\frac{S}{N}\right)_L = \frac{I_L(r_s, z_s)}{I_L(r_n, z_n)}$$
$$\overline{\left(\frac{S}{N}\right)_L} = \frac{\frac{1}{h}\int_0^h I_0(r_s, z_s, z_i)dz_i}{\frac{1}{h}\int_0^h I_0(r_n, z_n, z_i)dz_i} \tag{3.45}$$

where, h is the thickness of the water layer, I is the sound intensity, and the corner
marks "L" and "0" are the sounds received by the vertical array and the point receiver
respectively. Without losing generality, a constant factor is ignored, and the sound
intensity received by the point receiver can be expressed as:

$$I_0(r, z, z_i) = |\phi(r, z, z_i)|^2 \tag{3.46}$$

where, R is the sound field of point source in shallow water channel, which can be understood as being proportional to sound pressure. It has the following two expressions: normal mode expression and ray expression, i.e.:

$$\varphi_{\text{normal mode}} = \pi i \sum_m U_m(z) U_m(z_i) H_0^{(1)}(k_m r) = \sum_m \varphi_m \tag{3.47}$$

$$\varphi_{\text{ray}} \approx \sum_m \left(\frac{e^{jk_0 R_{m1}}}{R_{m1}} + V_s \frac{e^{jk_0 R_{m2}}}{R_{m2}} + V_b \frac{e^{jk_0 R_{m3}}}{R_{m3}} + V_s V_b \frac{e^{jk_0 R_{m4}}}{R_{m4}} \right) \cdot (V_s V_b)^m \tag{3.48}$$

Equations (3.47) and (2.14) are the same except for constant coefficients. $U_m(z)$, $U_m(z_i)$ are the source excitation function of the normalized mth normal mode wave and the depth distribution function of its amplitude. They have the same function form. k_m is the horizontal wave number of the m-th normal mode wave. V_b, V_s are the reflection coefficients of seabed and sea surface respectively. Equation (3.47) regards the sound field as the superposition of normal mode waves, and Eq. (3.48) regards the sound field as the sum of the contributions of point sources and a series of virtual sources. The contribution of each virtual source to the sound field follows the law of spherical waves, and the strength of the virtual source is weighted according to the reflection coefficient and the number of reflections. In Eq. (3.48), the wave number is k_0, which is an approximation, that is, the waveguide is regarded as a uniform waveguide, which is desirable for the near field, because in this case, the bending of the ray is not important, so the vertical variation of the sound velocity of the medium can be ignored. The signal from each virtual source to the receiving point and the corresponding multiple reflected ray have the same contribution to the sound field. The principle of virtual source is shown in Fig. 3.24, and we have:

$$R_{m1} = \sqrt{r^2 + (2mh + z_i - z)^2}$$
$$R_{m2} = \sqrt{r^2 + (2mh + z_i + z)^2}$$

Fig. 3.24 The diagram of shallow water sound field in the perspective of ray (sound field is regarded as the superposition of a series of virtual source sound fields)

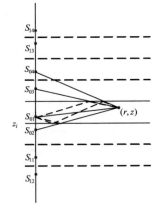

$$R_{m3} = \sqrt{r^2 + [2(m+1)h - z_i - z]^2}$$

$$R_{m4} = \sqrt{r^2 + [2(m+1)h - z_i + z]^2}$$

For the far field, the received sound intensity is:

$$I_0(r, z, z_i) = |\varphi_{\text{normal mode}}|^2 = \sum_m \psi_m \psi_m^* + \sum_j \sum_{\neq m} \psi_j \psi_m^* \tag{3.49}$$

where ψ_m is the wave function of the mth simple normal wave. When the depth is averaged on the right side of Eq. (3.49), the second term is usually far less than the first term, which can be omitted. If the depth is averaged, the average sound intensity is as follows:

$$\overline{I_0(r, z)} = \frac{1}{h} \int_0^h I_0(r, z, z_i) dz_i \approx \frac{1}{h} \int_0^h \sum_m |\psi_m|^2 dz_i \tag{3.50}$$

By substituting Eq. (3.47) into Eq. (3.50), we can use the asymptotic expansion of Hankel function to calculate the far-field sound intensity

$$\int_0^h U_m^2(z_i) dz_i \approx 1$$
$$k_m = \mu_m + j\beta_m \tag{3.51}$$

Then Eq. (3.50) becomes:

$$\overline{I_0(r, z)} \approx (\frac{2\pi}{rh}) \sum_m U_m^2(z) \frac{e^{-2\beta_m r}}{|k_m|} \tag{3.52}$$

To sum up, the average sound field of point receiver is as follows:

$$\overline{I_0(r, z)} = \begin{cases} \overline{|\varphi_{\text{normal mode}}|^2} = (\frac{2\pi}{rh}) \sum_m U_m^2(z) \frac{e^{-2\beta_m r}}{|k_m|} \\ |\varphi_{\text{ray}}|^2 = \frac{A(r,z)}{R^2(r,z)} \end{cases} \tag{3.53}$$

Equation (3.53) is the result of the ray concept. In homogeneous medium, the sound field of point source follows the law of spherical wave. In layered medium, the average sound field of point source along the depth is considered that only one abnormal transmission factor $A(r, z)$ needs to be corrected for the law of spherical wave. $R(r, z)$ is the average equivalent distance. When $r > h$, there is $R(r, z) = r$. Because the expressions of A and R are not used in the following analysis, we will discuss them later.

If the weight function of the vertical linear array is $\omega(z_i)$, the output of the linear array is:

$$I_L(r, z) = \left| \int_0^h \omega(z_i) \sum_m \pi i U_m(z) U_m(z_i) H_0^{(1)}(k_m r) \, dz_i \right|^2 \tag{3.54}$$

In Eq. (3.54), if $z = z_i$ and $r = r_s$ is taken, $z = z_n$ and $r = r_n$ can get the target signal output by linear array and the sound intensity of interference respectively. Substituting Eqs. (3.52)–(3.54) into Eq. (3.44), we have:

$$G = \frac{r_n A(r_n, z_n)}{R^2(r_n, z_n) \left(\frac{2\pi}{rh} \right) \sum_m U_m^2(z_s) \frac{e^{-2\beta_m r_s}}{|k_m|}} \cdot F(r_s, z_s, r_n, z_n, \omega) \tag{3.55}$$

where,

$$F(r_s, z_s, r_n, z_n, \omega) = \frac{\left| \sum_m U_m(z_s) f(m, \omega) e^{i\mu_m r_s - \beta_m r_s} / \sqrt{|k_m|} \right|^2}{\left| \sum_m U_m(z_n) f(m, \omega) e^{i\mu_m r_n - \beta_m r_n} / \sqrt{|k_m|} \right|^2} \tag{3.56}$$

$$f(m, \omega) = \int_0^h \omega(z_i) U_m(z_i) \, dz_i \tag{3.57}$$

The weight function is expanded into the series of eigenfunctions, i.e.,

$$\omega(z_i) = \sum_m A_m U_m(z_i)$$

where,

$$A_m = \int_0^\infty \omega(z_i) U_m(z_i) dz_i$$

$$\approx \int_0^h \omega(z_i) U_m(z_i) dz_i = f(m, \omega) \tag{3.58}$$

Those normal waves with lower number are waveguide normal waves, most of their energy transmits in the water layer, and there is only a little energy in $z > h$ space, so it is allowed to take an approximation to Eq. (3.58).

The Eq. (3.58) is substituted into formula (3.56), thus we have:

$$F(r_s, z_s, r_n, z_n, \omega) = \frac{\left| \sum_m U_m(z_s) \frac{A_m}{\sqrt{k_m}} e^{i\mu_m r_s - \beta_m r_s} \right|^2}{\left| \sum_m U_m(z_n) \frac{A_m}{\sqrt{k_m}} e^{i\mu_m r_n - \beta_m r_n} \right|^2} \tag{3.59}$$

The space gain of vertical linear array is estimated for different cases.

(1) When r_s, z_s, r_n, z_n is known a priori, in order to make the vertical linear array match the channel to get the maximum spatial gain, we only need to adjust the weight function $\omega(z_i)$, that is, adjust A_m to make $F(r_s, z_s, r_n, z_n, \omega)$ reach the maximum. We are not going to discuss the mathematical problems related to the selection method of the best weight function, but to selecting the weight function to match the signal field, and we obtain:

$$A_m = \begin{cases} U_m^*(z_s)e^{-i\mu_m r_s - \beta_m r_s} & \text{when } m < m_{\text{eff}}(r_s) \\ 0 & \text{others} \end{cases}$$

Substituting Eq. (3.59), we obtain:

$$F(r_s, z_s, r_n, z_n, \omega) = \frac{\left| \sum\limits_{m}^{m_{eff}(r_s)} |U_m(z_s)|^2 \frac{1}{\sqrt{k_m}} e^{-2\beta_m r_s} \right|^2}{\left| \sum\limits_{m}^{m_{eff}(r_s)} U_m(z_n)U_m^*(z_s) \frac{1}{\sqrt{k_m}} e^{-i\mu_m(r_s - r_n) - \beta_m(r_s + r_n)} \right|^2} \qquad (3.60)$$

where $m_{eff}(r_s)$ is defined as:

$$2(\beta_{m_{eff}} - \beta_1)r_s = 1$$

β_m is the attenuation factor of the m-th normal mode wave. The meaning of Eq. (3.60) is that at the distance r_s, the amplitude of the m_{eff} normal mode wave is $1/e$ of the amplitude of the first normal mode wave. Therefore, for a given distance r_s, there will be only m_{eff} low order normal modes which have important contributions to the sound field.

It is recorded as:

$$M(r_s) = \sum_{m=1}^{m_{eff}(r_s)} e^{-2(\beta_m - \beta_1)r_s} \qquad (3.61)$$

It proves:

$$m_{eff}(r_s) \approx M(r_s) = \sum_{m=1}^{m_{eff}(r_s)} e^{-2(\beta_m - \beta_1)r_s} \qquad (3.62)$$

The "rectangle method" can be used to approximate the sum of Eq. (3.61), thus proving the above formula. Each component in the summation sign in Eq. (3.61) is plotted according to the serial number m, as shown in Fig. 3.25. The summation value

Fig. 3.25 Schematic
diagram of relative amplitude
of normal mode wave

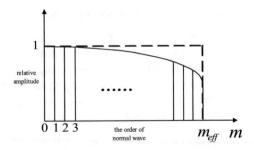

of Eq. (3.62) is the area of the shadow area, which can be approximately replaced
by the rectangular area surrounded by the dotted line and coordinates. It is easy to
know that the rectangular area is equal to m_{eff}, so Eq. (3.62) is held approximately.

In Eq. (3.60), because the signal field has been matched, the molecules are superimposed in-phase, while the denominator is not superimposed in-phase. When the
order of normal wave is not very small, it is close to the result of superposition by
energy. In the denominator, r_n can be ignored compared with r_s, and the sound field
is smoothed according to the depth, so as to blur the depth information. There is
$U_m^2 \propto \frac{1}{h^2}$ [5]. Note Eq. (3.62), and take $k_m \approx k_0$, then Eq. (3.55) becomes:

$$\overline{G} \approx \frac{r_n A(r_n, z_n)}{R^2(r_n, z_n)\left(\frac{2\pi}{k_0 h^2}\right)m_{eff}(r_s)} \cdot m_{eff}(r_s) = \left(\frac{h^2}{\lambda r_n}\right) A(r_n, z_n) \qquad (3.63)$$

where, λ—wave length.

(2) If the vertical array only matches the m-th normal mode wave, no prior
knowledge about r_s, z_s, r_n, z_n is needed, and the weight function is the same as
the amplitude distribution function of the m-th normal mode wave, i.e.:

$$w(z_i) = U_m(z_i)$$

That is:

$$A_k = \begin{cases} A_m & k = m; \\ 0 & others \end{cases}$$

It can be seen from Eq. (3.63):

$$I_L(, r, z) = \left| \frac{2\pi}{r} U_m(z) \frac{1}{\sqrt{k_m}} e^{j\mu_m r - \beta_m r} \right|^2 \qquad (3.64)$$

Substituting into Eq. (3.55), we obtain

$$G_m = \frac{r_n U_m^2(z_s) A_m e^{-2\beta_m(r_s-r_n)}}{R^2(r_n, z_n) \cdot U_m^2(z_n)\left(\frac{2\pi}{k_0 h}\right) \sum_m U_m^2(z_s) e^{-2\beta_m r_s}} \tag{3.65}$$

If the sound field is smoothed according to the depth coordinate, the average gain can be obtained as follows:

$$\overline{G} \cong \left(\frac{h^2}{\lambda r_n}\right) \frac{A}{m_{eff}(r_s)} e^{-2(\beta_m-\beta_1)r_s} \tag{3.66}$$

(3) If only the first normal mode wave is matched, then there are:

$$\sum_m U_m^2(z_s) e^{-2\beta_m r_s} \approx U_1^2(z_s) e^{-2\beta_1 r_s}$$

The above formula is approximately true at a long distance, that is, when the distance is very long, only the first normal wave makes an important contribution to the sound field. Therefore:

$$\overline{G} = \left(\frac{h}{\lambda r_n}\right) \cdot \frac{A}{U_1^2(z_n)} \tag{3.67}$$

The gain can be estimated by Eqs. (3.63), (3.66) and (3.67) only by calculating normal mode wave and smooth sound field [5]. As an example, if h = 60 m, r_n = 100 m, r_s = 40 km, in the case of sand bottom, when λ = 1 m, the spatial gain of vertical linear array matching with the first normal mode wave is about 15 dB.

3.9 Interference Structure of Low Frequency Short Range Sound Field

The considerable wave growth of low-frequency sound wave results in strong coherence of sound field. This is because the scale of the random inhomogeneity of the environment is smaller than that of the wavelength, and the ray is almost straight in the near distance, the influence of the sound velocity distribution and inhomogeneity on the sound field is small, and the coherence of the sound field is highlighted. Since the sound field has a stable interference structure, this kind of interference pattern of the sound field is called LOFAR diagram.

Low frequency and short-range sound field is of great significance to the application of underwater detonator fuse and target radiation noise characteristic measurement. This section provides the basic concepts.

Firstly, the ray acoustic method is used to predict the sound field, and the results are compared with the wave theory. The results show that as long as $f \geq 10\frac{c}{h}$ (h is the sea depth), the ray theory results have enough high approximation accuracy.

The ocean with the first layered media is investigated here, that is, the ocean interface is plane, and the seawater and seabed media are homogeneous fluids. The reflection coefficient of the sea surface is -1. The reflection coefficient of seafloor accords with the law of terminal reflection described in Sect. 2.4.

According to the view of ray acoustics, the sound field at the receiving point is the superposition of the direct sound and the multi-path reflected sound at the interface, which is equivalent to the superposition of the spherical waves emitted by the point source and a series of virtual sources at the receiving point. The geometric relationship between virtual source and point source is shown in Fig. 3.24. The sea surface coordinate is $z = 0$ and the seabed is $z = h$. The point source is located at $(0, z_0)$ and the receiving point is (r, z).

According to Eq. (3.48), $\phi(r, z)$ is:

$$\phi(r, z) = \sum_{l} \left(\frac{e^{jk_0 R_{l1}}}{R_{l1}} + V_s \frac{e^{jk_0 R_{l2}}}{R_{l2}} + V_b \frac{e^{jk_0 R_{l3}}}{R_{l3}} + V_s V_b \frac{e^{jk_0 R_{l4}}}{R_{l4}} \right) (V_s V_b)^l \quad (3.68)$$

where, the oblique distance $R_{\ell i}$ is:

$$R_{l1} = \sqrt{r^2 + (2\ell h + z_0 - z)^2}$$

$$R_{l2} = \sqrt{r^2 + (2\ell h + z_0 + z)^2}$$

$$R_{l3} = \sqrt{r^2 + [(2\ell + 1)h - z_0 - z]^2}$$

$$R_{l4} = \sqrt{r^2 + [(2\ell + 1)h - z_0 + z]^2} \quad (3.69)$$

V_s and V_b are the reflection coefficients of sea surface and seafloor respectively, and

$$V_s = -1 \quad (3.70)$$

$$V_b = \frac{m \sin \theta_{li} - \sqrt{n^2 - \cos^2 \theta_{li}}}{m \sin \theta_{li} + \sqrt{n^2 - \cos^2 \theta_{li}}}, i = 1, 2, 3, 4 \quad (3.71)$$

where, the acoustic refractive index $n = c_1/c_2$ and the density ratio $m = \rho_2/\rho_1$. The acoustic impedance of seawater medium and seabed medium is $\rho_1 c_1$ and $\rho_2 c_2$, respectively.

In Eq. (3.71), the sweep angle of ray θ_{li} is:

$$\begin{array}{l} \theta_{l1} = tg^{-1} \frac{2lh + z_0 - z}{r} \\ \theta_{l2} = tg^{-1} \frac{2lh + z_0 + z}{r} \\ \theta_{l3} = tg^{-1} \frac{2(l+1)h - z_0 - z}{r} \\ \theta_{l4} = tg^{-1} \frac{2(l+1)h - z_0 + z}{r} \end{array}, l = 0, 1, 2 \cdots \cdots \quad (3.72)$$

Table 3.5 Marine channel parameters for calculation

Numbers	f (Hz)	$h(m)$	$z_0(m)$	$z(m)$	kh
1	20	47	43.4	1.5	3.94
2	200	47	43.4	1.5	39.37

According to Eqs. (3.68)~(3.72), the ray theoretical sound field can be calculated. Equation (3.68) considers that the sound field at the receiving point is the superposition of spherical waves emitted by the sound source and a series of virtual sources.

According to the wave theory, the long-range sound field of point source in layered media can be expressed as normal mode wave (see Eq. (2.14)), and the contribution of side wave and normal mode wave with complex eigenvalue must be added in the short range. A detailed mathematical description will not be given here. Many software have been developed to calculate the short-range wave theory sound field. In this section, the numerical results are given and compared with the ray theory results (Table 3.5).

Although ray theory has a high-frequency approximation, Fig. 3.27 intuitively shows that it is in good agreement with the prediction results of wave theory when $f > 200$ Hz. For low-frequency sound waves, even in shallow water, ray theory can predict sound field with reasonable accuracy, and it can take advantage of its simple algorithm, which should be paid attention to. Figures 3.26 and 3.28 confirm this view again. The former has a higher frequency of sound wave, so the consistency between ray and wave theory reaches good agreement, while the latter has a lower frequency, so the consistency between them is very poor. In a word, the approximation accuracy of ray theory is satisfactory when $f > 10\frac{c}{h}$ (Fig. 3.31).

Fig. 3.26 The comparison of prediction results of ray and wave methods (Channel parameters can be seen in Table 3.5, Type 2.)

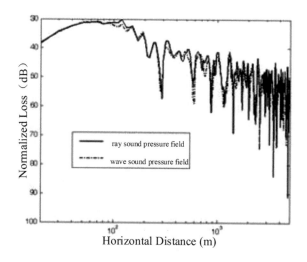

ray sound pressure field

wave sound pressure field

Fig. 3.27 The comparison
of prediction results of two
methods (range is 300 m, the
channel parameters can be
seen in Table 3.5)

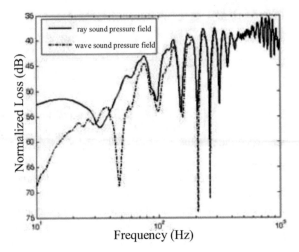

Fig. 3.28 The comparison
of prediction results of ray
and wave methods (The
channel parameters are
shown in table 3.5, type 1)

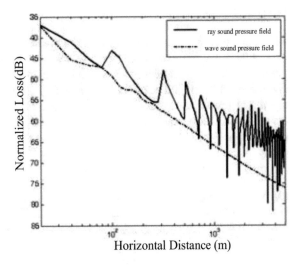

The above picture shows that the sound field of theoretical prediction is reasonably consistent with the experimental results, and the consistency of the two theories is also good.

The sea trial conditions in Fig. 3.29 are as follows: a surface ship is sailing in a straight line at a constant speed, the hydrophone is placed on the bottom of the sea with the sea depth of 47 m, and the average periodogram power spectrum of its output acoustic signal is analyzed. The power spectra obtained at different distances are combined into a time–frequency diagram, which is called LOFAR diagram. The horizontal coordinate in the figure represents time, which is converted into horizontal distance, namely distance/frequency diagram. In the figure, the horizontal distance is 0, which is the time when the target remains closest to the hydrophone.

Fig. 3.29 The sea trial results of short-range sound field

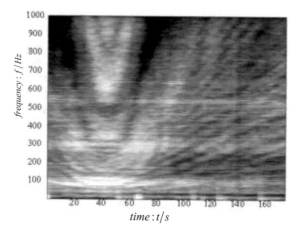

Fig. 3.30 The short range sound field predicted by wave theory (the conditions are the same as Fig. 3.29)

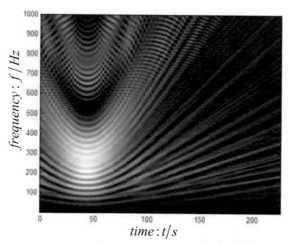

Fig. 3.31 The short range sound field predicted by ray theory (the conditions are the same as Figs. 3.29 and 3.30)

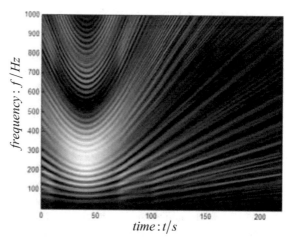

The clear interference fringes on LOFAR image show that the coherence of low frequency sound field is very strong. The ocean between the target sound source and the receiving point can be regarded as a filter, which transforms the source radiation spectrum into the receiving spectrum and distorts the spectrum of the source radiation signal. The dark fringes in the image are multi-path interference fringes; Bright fringes are enhanced interference fringes. If the target radiates a line spectrum sequence, a series of dark stripes will result in the defect of the line spectrum sequence; Even if the line spectrum sequence radiated by the target is stable, the received line spectrum sequence is incomplete and related to the geometric positions of the source and the receiver.

The interference structure of low frequency sound field has an important influence on the measurement of target radiated noise and the performance of underwater acoustic fuse.

3.10 Sound Pressure Time Reversal Mirror

The ideal channel can transmit information without distortion. However, the multi-path effect of the ocean and the inhomogeneous line of the medium will distort the time domain characteristics of the received signal (time delay extension), amplitude and phase fluctuations (frequency extension). In the past, the method to deal with the inhomogeneity of propagation medium is to know the prior knowledge of the inhomogeneous distribution of acoustic parameters in the medium space, and then to correct it. However, in many cases, it is very difficult or even impossible to obtain a detailed prior knowledge of the medium. How to eliminate the phase distortion, waveform distortion and image distortion caused by medium inhomogeneity is an important problem in focusing, detecting and acoustic imaging technology [6].

In order to solve these problems, in the late 1980s, the phase conjugation method originated from optics was introduced into acoustics, and then further developed into time reversal mirror method [7, 8]. Since Fink et al. drew the conclusion of focusing ability of time reversal array in ultrasound in 1989 [9], time reversal technology has become a hot spot in theoretical and experimental research of scientists [10, 11].

The focusing principle of time reversal mirror is described by one experiment. Figure 3.32 shows that W. A. Kuperman, William S. hodskiss and others conducted a sea trial of time reversal mirror (TRM) on the west coast of Italy in April 1996 [12].

In the experiment, a 77 m long SRA (source receive array) is placed in the sea with a depth of 125 m. The array consists of 20 transducers with the same frequency response. Another 90 m long hydrophone vertical receiving array (VRA) composed of 46 elements is located at the distance of 6.3 km from SRA. The point source (PS) is in the vertical plane composed of SRA and VRA. The test steps are as follows: ①. SRA receives a single frequency pulse with pulse width of 50 ms and 445 Hz from PS, and retransmits it after time reversal; ② VRA records the signal energy sent back by SRA at different depths; ③ Change the distance between PS and VRA, and record

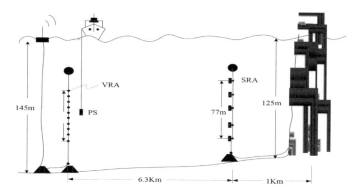

Fig. 3.32 The Phase conjugation test in 1996

the energy of VRA received signal at different distances; ④ The surface of energy of phase conjugate signal received by VRA with respect to distance and depth is given.

The results of the time reversal mirror sea trial displayed in Fig. 3.32 are shown in Fig. 3.33. Figure 3.33a shows the pulse waveform received by SRA when PS depth is 75 m; Figure 3.33b is the signal waveform received by VRA after the signal in Fig. 3.33a when being transmitted again after time reversal; Figure 3.33c, d are the spectrum of the waveform of Fig. 3.33a, b respectively.

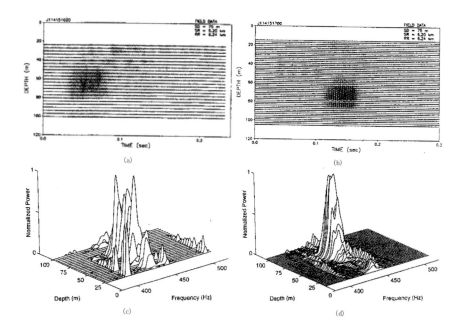

Fig. 3.33 The sea trial results of time reversal mirror

Fig. 3.34 Principle block diagram of sea trial of active sound pressure reversal mirror

The experimental results show that the energy of time reversal signal is the strongest near the sound source (PS). It is proved that the time reversal mirror (corresponding to the time domain time reversal mirror) can achieve time and space refocusing at the original sound source position in the marine environment; Compared with free space environment, ocean waveguide has better focusing effect.

The above sea trial process can be summarized as the flow chart in Fig. 3.34.

The following is about the basic principle of the time reversal method in the distance independent waveguide medium, which uses the channel multipath characteristics to achieve focusing.

In Fig. 3.32, the sound field of the J-th element in the SRA of the transceiver combined array at the point R from PS can be expressed by Green's function $G_\omega(R; z_j, z_{ps})$, where $|R|$ is the horizontal distance between PS and SRA; z_{ps} represents the depth of PS; z_j represents the depth of element j in SRA, $G_\omega(R; z_j, z_{ps})$ satisfies Helmholtz equation in the presence of sound source

$$\nabla^2 G_\omega(R; z_j, z_{ps}) + k^2(z_j)G_\omega(R; z_j, z_{ps}) = -\delta(R - r_{ps})\delta(z_j - z_{ps}) \quad (3.73)$$

The time-domain expression of the acoustic field at the element j can be obtained by Fourier synthesis:

$$P(R, z_j; t) = \int G_\omega(R; z_j, z_{ps})S(\omega)e^{-i\omega t}d\omega \quad (3.74)$$

where, $S(\omega)$ is the Fourier transform of the signal from the point source. For convenience, the sound pressure $P(R, z_j; t) \neq 0$ is set within $(0, \tau)$ and 0 at other times. Therefore, based on the pulse transmitting time, the time reversal signal of the j-th element of SRA can be expressed as $P(R, z_j; T - t)$, where $T > 2\tau$, which ensures the causality of the system. Thus, there are:

$$P(R, z_j; T - t) = \int G_\omega(R; z_j, z_{ps})S(\omega)e^{-i\omega(T-t)}d\omega$$

$$= \int \left[G_\omega^*(R; z_j, z_{ps})e^{i\omega T}S^*(\omega)\right]e^{-i\omega t}d\omega \quad (3.75)$$

Formula (3.75) describes the Fourier transform of the pulse signal received by the j-th element in SRA after being delayed and reversed, which reflects the corresponding relationship between time reversal in time domain and phase conjugation in frequency domain.

The time reversal signal is transmitted into the sound field again. At this time, the sound pressure $p_{pc}(r, z; t)$ at any observation point (r, z) in the sound field is:

$$P_{pc}(r, z; t) = \sum_{j=1}^{J} \int G_\omega(r, z, z_j) G_\omega^*(R, z_j; z_{ps}) e^{i\omega T} \times S^*(\omega) e^{-i\omega t} d\omega \quad (3.76)$$

Equation (3.76) shows that TRM can achieve spatial focusing.

For a known signal $s(t)$, if the signal spectrum is $S(j\omega)$, the transfer function $H(j\omega)$ of the matched filter should meet the following requirements:

$$H(j\omega) = kS^*(j\omega) e^{-i\omega t_0} \quad (3.77)$$

where, k is a constant. The equation shows that when the frequency characteristic of the matched filter is conjugate with the frequency spectrum of the input signal [13], the output signal-to-noise ratio of the filter reaches the maximum at t_0.

At the same time, the impulse response function ht of the matched filter can be:

$$h(t) = ks(t_0 - t) \quad (3.78)$$

The above formula shows that if the impulse response function of the filter is the image of the input signal $s(t)$ and shifts t_0 in time, then the filter can perform matched filtering on $s(t)$ and achieve the maximum signal-to-noise ratio at t_0.

Combined with Eq. (3.77) and Eq. (3.78), we analyzed Eq. (3.76). If $S(i\omega) = G_\omega(r, z, z_j)$, where j represents the j-th element in SRA, when $r = R, z = z_{ps}$, i.e. the position of sound source is observed, $G_\omega(r, z, z_j) = G(R, z_{ps}, z_j)$. As a linear network, the excitation and response in ocean channel have reciprocity, so $G(R, z_{ps}, z_j) = G(R, z_j, z_{ps})$. It can be concluded that the "$G_\omega^*(R, z_j; z_{ps}) e^{i\omega T}$" part in Eq. (3.76) is the transfer function of the matched filter of $G(R, z_j, z_{ps})$. In other words, the time reversal process in Fig. 3.34 forms a matched filtering process. The matched filtering here does not match the transmitted signal $s(t)$ of PS, but matches the sound channel I of acoustic transmission. Therefore, this matching process is called spatial (channel) matching. Another reason why we call it spatial focusing is that we assume that $r = R, z = z_{ps}$, which means that only in the position of the original sound source can the channel be matched to focus the signal energy after time reversal. The third reason of realizing spatial matched filtering with time reversal mirror is the contribution of each element of vertical array. The process of adding the results of all the elements is a process of spatial filtering. The addition of the signals of each channel enhances the focusing effect, which is similar to that of beamforming.

3.11　Vector Time Reversal Mirror

It is pointed out in Sect. 1.4 that the particle velocity field can be expressed by the negative gradient of the non directional sound pressure field. It can be seen from Eq. (1.15) that the vibration velocity field only changes the directional characteristics of the sound pressure field, but does not change the reciprocity and time characteristics of the sound pressure field. Therefore, the concept and method of time reversal can be applied to the vibration velocity field to form a "vector reversal mirror".

Figure 3.35 shows the block diagram of the vector reversal mirror.

The vector reversal mirror also reverses the received signal with the local interference and then passes through the channel again. When the two channels through which the signal passes are completely matched, the output has the maximum spatial gain, that is, when channel II and channel I are completely matched, that is, when the observation point (r_i, h_i) coincides with the sound source (r_0, h_0), the spatial gain reaches the maximum. In addition, a vibration velocity channel is added to the vector reversal mirror. From the analysis in the previous sections, it can be seen that the signal components in $z_p(t)$ and $z_V(t)$ are correlated, while the isotropic interference components are uncorrelated. Accordingly, cross correlator can be used as post processor to suppress isotropic interference and improve output SNR.

Therefore, the detection threshold of the vector reversal mirror using the vibration velocity information can be lower than that of the acoustic pressure time reversal mirror. This is because the vector reversal mirror is a practical spatiotemporal filter, which not only matches the channel space matching filter, but also matches the unknown target radiation signal waveform. Therefore, from the point of view of space–time matching, the vector reversal mirror should have good detection ability.

From the perspective of channel matching, the pressure reversal mirror and the vector reversal mirror are the same. This paper takes the pressure signal as an example to analyze the spatial correlation characteristics of the reversal mirror. Assuming that the target signal $x(t)$ is sent from the source $S(r_0, h_0)$ and reaches the receiving point R through the ocean multipath channel I (the impulse response function is represented by $h_p(\tau, r_0, h_0)$), the output signal $y_p(t)$ of the receiver can be:

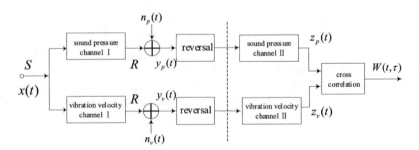

Fig. 3.35 Vector inversion frame

$$y_p(t) = x(t) * h_p(\tau, r_0, h_0) + n_p(t) \qquad (3.79)$$

where $n_p(t)$ is the noise component of the local interference, $*$ represents the convolution operation.

After time reversal of $y_p(t)$, the output $z_p(t)$ can be expressed as:

$$
\begin{aligned}
z_p(t) &= y_p(-t) * h_p(\tau, r, h) \\
&= x(-t) * h_p(-\tau, r_0, h_0) * h_p(\tau, r, h) + n_p(-t) * h_p(\tau, r, h) \\
&= x(-t) * \left\{ h_p(\tau, r, h) * h_p(-\tau, r_0, h_0) \right\} + n_p(-t) * h_p(\tau, r, h) \\
&= x(-t) * R_p(\varsigma, \rho, H) + n_p(-t) * h_p(\tau, r, h)
\end{aligned}
\qquad (3.80)
$$

where $R_p(\varsigma, \rho, H)$ is the correlation function between $h_p(\tau, r, h)$ and $h_p(\tau, r_0, h_0)$, $\rho = r_0 - r$, $H = h_0 - h$.

The signal components in $y_p(t)$ are convoluted twice, that is, they pass through the channel twice. When $r = r_0$, $h = h_0$, then $R_p(\varsigma, \rho, H) = R_p(\varsigma, 0, 0)$, that is, the channel is completely matched, and the corresponding correlation peak is $R_p(0, 0, 0) = \delta(\varsigma, \rho, H)/_{\varsigma=0, \rho=0, H=0}$, which means that the signal component in $z_p(t)$ comprehensively utilizes the energy of each channel to the ray, and it is also the in-phase superposition of the signal energy from each channel; The interference component in $y_p(t)$ only passes through the channel once, so the interference component in $z_p(t)$ is the interference superposition of the noise energy from each channel. Therefore, the time reversal mirror realizes the spatial matching of the signal, which is one of the ways to obtain the gain of the time reversal mirror technology.

The principle of the vibration velocity channel is the same as that of the sound pressure channel, and it will not be described here.

Figure 3.36 shows the three-dimensional rendering of TR spatial correlation characteristics, in which the target position is 1000 m horizontally and 30 m vertically.

Fig. 3.36 Three dimensional rendering of TR spatial correlation characteristics

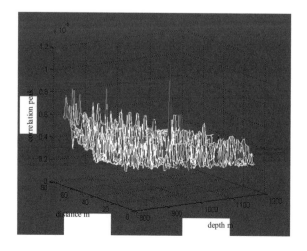

It can be seen from the three-dimensional effect picture of the spatial correlation characteristics of the inverted mirror that when the signal source position $S(r_0, h_0)$ and the observation position $A(r, h)$ coincide, the channels match completely, and the correlation peak of the two channels appears.

The space gain of vector sensor lies in the fact that it can make the two channels through which the signal passes match completely after it sends out the time reversal signal, and make comprehensive use of the energy of the arriving ray at the observation point; The interference and the acoustic signals from other observation points are suppressed.

So the space processing gain of vector sensor is defined as:

$$G = 20 \cdot \log_{10} \frac{\left| \int h(-t, r_0, h_0) * h(t, r_0, h_0) dt \right|}{\left| \int h(t, r_0, h_0) dt \right|} \tag{3.81}$$

In the above formula, $h_p(\tau, r_0, h_0)$ in the denominator represents the total channel through which the received signal passes without the reverse mirror technology; $h_p(-\tau, r_0, h_0) * h_p(\tau, r_0, h_0)$ in the molecule represents the total channel through which the received signal passes using the inverted mirror technology.

The following figure shows the simulation results of the spatial processing gain of the vector reversal mirror with the target between 800 and 1200 m, in which the observation point coincides with the target position, and the target and the receiver are in 20 m underwater.

Comparing Figs. 3.37 and 3.38, it can be seen that the larger the processing bandwidth is, the larger the spatial gain will be, because the wider the channel bandwidth is, the more frequency components of the signal will pass through, and the more components will be available at the observation point. This is particularly evident in the direction of the V_r.

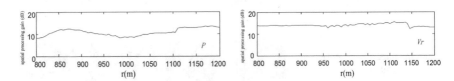

Fig. 3.37 TR processing gain (operating band 2 kHz $\sim$ 7 kHz)

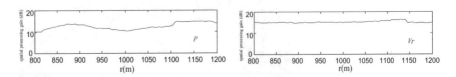

Fig. 3.38 TR processing gain (operating band 200 Hz $\sim$ 1 kHz)

References

1. Rihaczek AW. Principles of high-resolution radar. Norwood, MA: Artech House; 1996.
2. Hui JY, Wang LS. Adaptive matching filter and adaptive correlator. Underwater acoustic communication; 1986.
3. Tolstoy I, Clay CS. Ocean acoustic. McGraw-Hill Book Company; 1977.
4. Wang DZ, Shang EC. Underwater acoustics. Beijing: Science Press; 1984.
5. Chen G, Xu JH. Coherence measurement of shallow sea channel—pulse correlation method. Acta Acustica (Chinese version); 1983.
6. Wei W, Study on adaptive focusing performance of ultrasonic time reversal method. Ph.D. Thesis
7. Brysev AP, et al. Phase conjugation of acoustic beams. Sov Phys Acoust. 1987;32:408–10.
8. Burkin F, et al. Acoustic analogues of nonlinear optics phenomena. Sov Phys Acoust. 1987;29(7):607.
9. Fink M, et al. Self focusing in inhomogeneous media with "time reversal" acoustic mirrors. Proc IEEE Ultrason, Symp. 1989;2:681–6.
10. Fink M, et al. Time-reversed acoustics. Rep Prog, Phys. 2000;63:1993–5.
11. Fink M, et al. Acoustic time-reversal mirrors. Inverse Probl. 2001;17:1–38.
12. Hee CS, Kuperman WA, Hodgkiss WS. A time-reversal mirror with variable range focusing. J Acoust Soc Am. 1998;103:3234–40.
13. Zhu H. Stochastic signal analysis. Beijing Institute of Technology Press; 1995

Chapter 4
Theoretical Basis of Random Time-Varying Space-Varying Channel

In the above chapters, the sea water medium and its boundary are regarded as linear time invariant, that is, deterministic. Also, the ocean acoustic channel is regarded as a deterministic linear time invariant filter or a deterministic time–space filter, which is called coherent multipath channel model. The coherent multipath channel not only transforms the energy of the sound source radiation signal, but also transforms the waveform and forms the spatial interference pattern, resulting in complexity of underwater acoustic signal processing.

The main part of the multi-path energy in the ocean is coherent. The experimental results show that the transverse correlation radius of single frequency continuous wave is very large, serving as an evidence to support the above argument. The sound field in the ocean has a relatively stable interference pattern, and a stable convergence region can be seen even at a long distance, which fully shows that the energy of multi-path arrival is coherent. In Chap. 5, it will prove that the ocean channel can be regarded as a slow time-varying coherent multipath channel in most sonar applications. However, in other important applications, the stochastic process of ocean channel must be considered, and only by understanding the various stochastic phenomena in sound propagation can we deeply understand the nature of coherent multipath channel model and its application limitations. In this chapter, the ocean is regarded as a random time-varying and space-varying filter, which transforms the sound source radiation signal randomly.

It is reasonable to assume that the ocean medium (including boundary and target) between the sound source and the receiver is a linear random time-varying and space-varying filter, unless the sound wave is strong enough to make the acoustic nonlinearity occur.

In this chapter, the basic concepts of some random processes in sound propagation are briefly introduce, while the in-depth study of these random processes is beyond the scope of this book.

This chapter will focus on the basic description methods of random channels, their system functions, and their physical meanings.

© Harbin Engineering University Press 2022
J. Hui and X. Sheng, *Underwater Acoustic Channel*,
https://doi.org/10.1007/978-981-19-0774-6_4

4.1 General Concept and Description of Random Sound Field

There are many kinds of random inhomogeneities in the ocean. The sea surface has random time-varying and spatial-varying uneven waves; The seawater medium is not uniform, and there are random water masses in the seawater, which are called temperature microstructure; There are random swimming fish, shrimp and plankton groups in the seawater; The topography and acoustic characteristics of the seafloor are also random and non-uniform; There are also random internal waves and tides in the ocean.

How does the inhomogeneity in the ocean affect the sound field? In the process of sound signal transmission in the ocean, the inhomogeneity of sea water medium and its boundary will cause random sound scattering. The sound energy back scattered to the receiver near the sound source will be superimposed to form the so-called reverberation. The energy of forward scattering will cause the fluctuation of the amplitude and waveform of the received signal. It is also important that the wave will lead to the random change of the relative phase relationship of the multi-path arrival signal, which will lead to the change of the spatial pattern of the sound field interference and the random change of the received signal. The statistical characteristics of the received signal depend not only on the incoherent energy component of the scattered sound wave, but also on the interference state of the coherent component, that is, the spatial structure of the deterministic sound field will also play an important role in the statistical characteristics of the signal.

If the parameters of the medium do not change with time and the position of the sound source and receiver is fixed, then the received signal is stable. In this case, only when the sound source and the receiver make relative movement, the received signal changes. The movement of sound source or receiver will also cause the fluctuation of sound signal, which is of great significance in sonar technology.

Even if the sound source sends out the same signal at different times, the signal received at a fixed receiving point will also change with time because the seawater medium and its boundary change with time, which is called acoustic signal fluctuation. In this case, the spatiotemporal statistical characteristics of the received signal can describe the characteristics of the random sound field. A general concept describing a random sound field will be introduced in the following part.

The total sound pressure $P(\boldsymbol{r}, t)$ received at a certain point in the ocean can be divided into two parts: deterministic sound field (coherent component) and random sound field (incoherent component).

$$P(\boldsymbol{r}, t) = p_0(\boldsymbol{r}, t) + p(\boldsymbol{r}, t) \tag{4.1}$$

The definite component of sound field is the average component of sound pressure,

$$p_0(\boldsymbol{r}, t) = \langle P(\boldsymbol{r}, t) \rangle \tag{4.2}$$

where $\langle \cdot \rangle$ is the ensemble average. $p_0(\mathbf{r}, t)$ is called deterministic sound field, or coherent component of sound field, also average sound field in some literatures; $p(\mathbf{r}, t)$ is called random sound field or incoherent component of sound field. The deterministic sound field $p_0(\mathbf{r}, t)$ is determined by the average sound velocity distribution and the average boundary characteristics, which determine the shadow area, the convergence area, and the spatial interference pattern of the sound field. In Chaps. 2 and 3, some problems about the deterministic sound field are discussed. The random component of sound field is produced by scattering and refraction of random inhomogeneous body, which should be described by statistical method.

In order to describe the random process $p(\mathbf{r}, t)$ completely, we need to use the infinite multidimensional distribution function. Fortunately, from an engineering point of view, only a few characteristic statistics can guide most engineering practices. Two commonly used statistical characteristics are mean $\langle p(\mathbf{r}, t) \rangle$ and correlation function $R_p(\mathbf{r}_1, t_1; \mathbf{r}_2, t_2)$.

$$R_p(\mathbf{r}_1, t_1; \mathbf{r}_2, t_2) = \langle [p(\mathbf{r}_1, t_1) - \langle p(\mathbf{r}_1, t_1) \rangle] \cdot [p(\mathbf{r}_2, t_2) - \langle p(\mathbf{r}_2, t_2) \rangle] \rangle \quad (4.3)$$

Equation (4.3) represents the cross-correlation function of signals received by two separate receiving points at different times, which is called the spatiotemporal correlation function of sound field.

According to Eq. (4.1), the mean value of $p(\mathbf{r}, t)$ of the random component of sound pressure is zero, so $\langle p(\mathbf{r}, t) \rangle = 0$, and Eq. (4.3) can be simplified as:

$$R_p(\mathbf{r}_1, t_1; \mathbf{r}_2, t_2) = \langle p(\mathbf{r}_1, t_1) \cdot p(\mathbf{r}_2, t_2) \rangle \quad (4.4)$$

Equation (4.4) is the cross-correlation function of general process. If the correlation function does not depend on absolute time but only on time difference $\tau = t_2 - t_1$, then the process is stationary in time domain; If the cross-correlation function does not depend on $\mathbf{r}$ but only on the difference $\rho = |\mathbf{r}_2 - \mathbf{r}_1|$, then such a random field is called spatially uniform random field or it is spatially stationary. Therefore, for spatiotemporal stationary random fields, the correlation function is as follows:

$$R_p(\mathbf{r}_1, t_1; \mathbf{r}_2, t_2) = R_p(\rho, \tau) = \langle p(\mathbf{r}, t) p(\mathbf{r} + \rho; t + \tau) \rangle \quad (4.5)$$

Generally, stationary stochastic processes in time domain are ergodic, and ensemble average of ergodic processes can be replaced by time average

$$R_p(\rho, \tau) = \overline{p(\mathbf{r}, t) p(\mathbf{r} + \rho, t + \tau)} \quad (4.6)$$

Sometimes we can distinguish the deterministic component from the random incoherent component, but generally we can't, so we can only study the correlation function of the total sound field, i.e.,

$$R_p(\rho, \tau) = \langle [p_0(\mathbf{r}, t) + p(\mathbf{r}, t)][p_0(\mathbf{r} + \rho, t + \tau) + p(\mathbf{r} + \rho, t + \tau)] \rangle$$

If it can be reasonably assumed that the deterministic component and the incoherent random component are independent of each other, then the average value of the cross terms in the above formula is zero. Therefore, we have:

$$R_p(\rho, \tau) = \langle p_0(\mathbf{r}, t)p_0(\mathbf{r} + \rho, t + \tau) \rangle + \langle p(\mathbf{r}, t)p(\mathbf{r} + \rho, t + \tau) \rangle \qquad (4.7)$$

In Eq. (4.7), the first term on the right is the coherent component. If $\rho = 0$, we can get the autocorrelation function as follows:

$$R_P(0, \tau) = \langle p_0(\mathbf{r}, t)p_0(\mathbf{r}, t + \tau) \rangle + \langle p(\mathbf{r}, t)p(\mathbf{r}, t + \tau) \rangle \qquad (4.8)$$

In the same way:

$$R_p(0, \tau) = \langle p(\mathbf{r}, t)p(\mathbf{r}, t + \tau) \rangle \qquad (4.9)$$

$R_p(0, 0)$ is called mean square deviation of fluctuation in sound field, that is, $R_p(0, 0) = \langle |p(\mathbf{r}, t)|^2 \rangle$ represents the energy of incoherent components of sound field, and $\langle |p_0(\mathbf{r}, t)|^2 \rangle$ represents the energy of deterministic sound field.

The fluctuation rate of acoustic signal is defined as:

$$\eta_p = \langle |p|^2 \rangle / \langle |p_0|^2 \rangle \qquad (4.10)$$

The fluctuation rate of sound field is equal to the ratio of the sound intensity of incoherent component to that of coherent component. Or the coherence coefficient of the sound field is defined as the ratio of the sound intensity of the coherent component to the total sound intensity, so the coherent coefficient is:

$$\xi = \frac{\langle |p_0|^2 \rangle}{\langle |p_0|^2 \rangle + \langle |p|^2 \rangle} = \frac{1}{1 + \eta_p} \qquad (4.11)$$

The coherence coefficient is used to describe the degree of coherence of sound field.

The correlation function is the most important one among the above statistics, as a special kind of function.

Its basic characteristics are: the correlation function should satisfy the general trend that $R(\rho, \tau) \leq R(0, 0)$, and the increase of $R(\rho, \tau)$ and τ is gradually decreasing, and it is the even function of ρ and τ. The correlation scale in time and space is called correlation radius, showing how fast the statistical correlation of sound field weakens. The correlation radius is equal to the spatial distance of the correlation function from $\rho = 0$ to 0.5 (or 1/E). The temporal correlation radius can be defined similarly.

4.2 Acoustic Signal Fluctuation

Due to the random inhomogeneous scattering and refraction of seawater medium and its interface, the signal waveform at the receiving point changes, and it is called acoustic signal fluctuation. Experiments show that the total received signal can be regarded as the superposition of coherent and incoherent components.

If the sound source radiates a single frequency harmonic signal, the time-varying nonuniformity leads to the broadening of the spectrum of the received signal (usually in the order of 0.1 Hz), and the fluctuation of the amplitude and phase of the signal. The total sound field can be written as:

$$P(t) = [E_0 + u(t)]\cos(\omega_0 t + \phi_0) - v(t)\sin(\omega_0 t + \phi_0) \qquad (4.12)$$

where E_0 is the amplitude of coherent signal component, ϕ_0 is the phase of coherent component, ω_0 is the angular frequency of acoustic source radiation signal, $u(t)$ and $v(t)$ represents two orthogonal components of incoherent scattering component, their mean value is zero and variance is σ.

Equation (4.12) can also be expressed as:

$$P(t) = E(t)\cos[\omega_0 t + \phi_0 + \Delta\phi(t)] \qquad (4.12a)$$

where

$$E(t) = \{[E_0 + u(t)]^2 + v^2(t)\}^{1/2}$$
$$\Delta\phi(t) = arctg\{v(t)/[E_0 + u(t)]\} \qquad (4.13)$$

Equation (4.13) is easier to see the fluctuation of amplitude and phase of total signal caused by scattering component.

Assuming that the $u(t)$ and $v(t)$ have a normal distribution, the amplitude $E(t)$ of the received signal satisfies the generalized Rayleigh distribution [1], and its distribution function $F(E)$ is:

$$F(E) = \left(\frac{E}{\sigma^2}\right)\exp[1 - (E^2 + E_0^2)/2\sigma^2]I_0(E_0 E/\sigma^2), \ E \geq 0$$
$$F(E) = 0, \ E < 0 \qquad (4.14)$$

where I_0 is a Bessel function of order zero of imaginary argument.

It can be proved from Eq. (4.14) that when the fluctuation is small (that is, when E_0/σ is large), the amplitude distribution of the signal is close to the Gaussian distribution with E_0 as the mean value; When the fluctuation is large ($E_0/\sigma \to 0$), the amplitude distribution of the total sound field is close to Rayleigh distribution. The

Fig. 4.1 The generalized
Rayleigh distribution

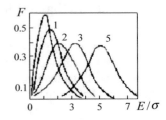

Fig. 4.2 Relation between
amplitude fluctuation rate of
direct sound and distance

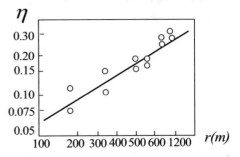

experiment supports the above conclusion. The generalized Rayleigh distribution is shown in Fig. 4.1.

In Fig. 4.1, Number 1 to 5 indicate the value of E_0/σ. The experimental results of the relationship between the amplitude fluctuation rate of direct sound and distance are shown in Fig. 4.2 [2]. The amplitude fluctuation of the signal is a common phenomenon, which is usually larger in the sound and shadow area and smaller in the bright area. The average period of amplitude fluctuation is about 10 s, and the period of amplitude fluctuation generated by internal wave is more than 1 min.

The phase fluctuation of acoustic signal increases with the increase of amplitude fluctuation. Generally, the variance of phase fluctuation is within tens of degrees, and larger phase fluctuation can be observed, but the fluctuation is very slow. For example, the sound transmission experiment [3] across the Florida Strait conducted by the Americans shows that for a 420 Hz sound wave, the distance is 40 n miles, the phase change of the received signal is not more than 100° in the range of 0.5–1.0 h, and the phase change is not more than 360° in 4 h. The acoustic phase changes very slowly, so the channel can be regarded as a slow time-varying coherent multipath channel, which is the physical basis for the application of adaptive coherent processing in underwater acoustics.

The influence of signal amplitude fluctuation on sonar system cannot be ignored. For threshold detector, signal fluctuation increases the probability of missing report when the signal clutter ratio is large, and increases the detection probability when the signal clutter ratio is small. But the latter has no practical significance, because the detection probability is very low currently, which is far from meeting the requirements of correct detection. For the time measurement system with simple threshold detector, the fluctuation of signal amplitude also leads to the increase of time measurement error. In underwater acoustic communication, signal fluctuation is also known as signal fading, which leads to either good or bad communication quality.

4.3   System Function of Time Varying Channel

The received signal is composed of deterministic components and incoherent components, according to which the ocean between the sound source and the receiving point is regarded as two parallel filters, as shown in Fig. 4.3. In this chapter, we will study the system function and characteristics of incoherent channel filter, as well as the influence on sonar signal processor.

This section describes the system functions of time-varying filters.

In Chap. 3, two system functions of time invariant filter are introduced: impulse response function $h(\tau)$ and transfer function $H(f)$. The two system functions are not independent of each other. Any one of them can completely determine the characteristics of the system. They are Fourier transforms. The output waveform of the filter is the convolution of the input waveform and the impulse response function, and the output spectrum of the filter is the product of the input signal spectrum and the transfer function.

Now let's define the system function of time-varying channel. Two basic input signals are used to measure the characteristics of time-varying networks, which are function δ and complex harmonic function. In underwater acoustic transmission experiments, two kinds of signals are often used to measure the characteristics of ocean acoustic channel: explosion pulse and continuous CW sound wave, which are equivalent to the above two basic signals respectively. As shown in Fig. 4.4, when the input of the system applies pulse before time t, the response value of the output of the time-varying network at time t is recorded as $h(\tau, t)$; $h(\tau, t)$ is called the impulse response function of the network. When the signal $e^{j2\pi ft}$ is added to the input of the system, the output of the system is recorded as $H(f, t)e^{j2\pi ft}$; $H(f, t)$ is called the transfer function of the network.

Both linear time invariant network and linear time-varying network follow the superposition principle. Following the discussion in Sect. 3.1, we can see that the

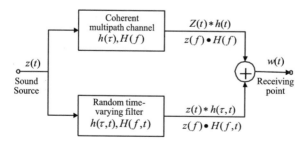

Fig. 4.3   Channel models of coherent and incoherent filters

Fig. 4.4 The system function of time-varying channel

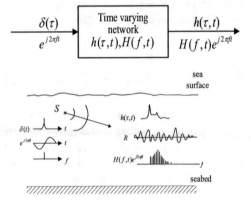

following expressions are true:

$$H(f, t) = \int h(\tau, t)e^{-j2\pi f\tau}\mathrm{d}\tau \tag{4.15}$$

$$h(\tau, t) = \int H(f, t)e^{j2\pi f\tau}\mathrm{d}f \tag{4.16}$$

$$w(t) = \int z(t - \tau)h(\tau, t)\mathrm{d}\tau \tag{4.17}$$

$$w(t) = \int Z(f)H(f, t)e^{j2\pi ft}\mathrm{d}f \tag{4.18}$$

where $z(t)$ is the input signal of the network and $w(t)$ is the output signal of the time-varying network.

Different from the time invariant network, when a complex harmonic signal $e^{j2\pi ft}$ is added to the input of the time variant network, the output of the network is $H(f, t)e^{j2\pi ft}$. Generally speaking, it is not a simple harmonic signal, but the one with amplitude and phase modulation. In other words, the response of the time-varying network to the continuous sinusoidal input signal is the signal modulated by the subcarrier frequency waveform $H(f, t)$. This phenomenon is called spectrum spread, which is a unique phenomenon in time-varying networks. The impulse response function $h(\tau, t)$ of time-varying network has two time variables, t is called absolute time, τ is called delay time. $h(\tau, t)$ can be understood as the instantaneous value of the network output at t, and $t + \tau$ time.

In order to intuitively understand the physical meaning of system functions of $h(\tau, t)$ and $H(f, t)$, we draw Fig. 4.5, and hope it can be helpful to readers. When a sinusoidal signal is added to the network input, the waveform and spectrum are shown in Fig. 4.5a, and the output spectrum of the time-varying network is no longer a single line spectrum $W(f)$; Figure 4.5b illustrates the input band limited white noise

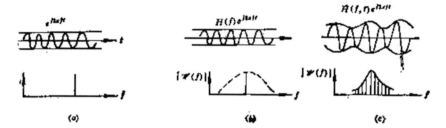

Fig. 4.5 Function diagram of time-varying network system. **a** Input signal waveform and spectrum.
b Time invariant network output waveform. **c** Time varying network output waveform

pulse sequence. The output spectrum of the time-varying network is stable, and the
output spectrum of the time-varying network will change with time t. Different from
time invariant networks, the response of time-varying networks to each pulse δ is
time-varying.

Both time-varying channels and time invariant channels spread the waveform of
the input signal in time. The unique feature of time-varying channels is to broaden
the spectrum of the input signal. The inhomogeneity in the ocean is moving, which
leads to Doppler spectrum broadening in the process of signal transmission. Making
Doppler spectrum analysis of the output of the time-varying network, we can better
understand the spectrum broadening behavior of the channel. According to Fig. 4.6,
we get two new system functions:

$$B(f, \phi) = \int H(f, t)e^{-j2\pi\phi t}\,dt \tag{4.19}$$

$$s(\tau, \phi) = \int h(\tau, t)e^{-j2\pi\phi t}\,dt \tag{4.20}$$

f is called the absolute frequency and ϕ is called the Doppler frequency. $B(f, \phi)$
is called dual frequency function. When the input frequency is a sinusoidal signal
of f, the complex amplitude of the sinusoidal component of the network output
frequency of $(f + \phi)$ is $B(f, \phi)$. From Eq. (4.20), we can know that $s(\tau, \phi)$ is
the time-varying spectrum of impulse response function, which describes the joint
expansion characteristics of channel in time domain and frequency domain, so it is
called expansion function.

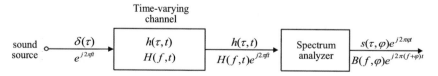

Fig. 4.6 Illustration of the four channel time-varying system functions

Fig. 4.7 The relationship diagram of four system functions

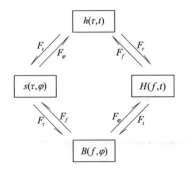

To sum up, time-varying networks have four system functions, any of which can completely determine the performance of the system. On certain aspect, they can describe the channel more intuitively and conveniently. The four system functions are Fourier transform pairs, and their mutual transformation relationship and corresponding variable pairs are shown in Fig. 4.7.

Readers can prove that the response of time-varying channel to any input waveform is:

$$w(t) = \int h(\tau, t)z(t-\tau)d\tau = \int H(f,t)Z(f)e^{j2\pi ft}df$$

$$= \iint s(\tau, \phi)z(t-\tau)e^{j2\pi\phi t}d\phi d\tau = \iint B(f, \phi)Z(f)e^{j2\pi(f+\phi)}d\phi df \quad (4.21)$$

where $Z(f)$ denotes frequency spectrum of signal $z(t)$.

The time invariant system can be regarded as a special case of time-varying system. At this time, the system function actually is degenerated into two, including:

$$\left.\begin{array}{l} h(\tau,\ t) = h(\tau), \quad s(\tau,\ \phi) = h(\tau) \cdot \delta(\phi) \\ H(f,t) = H(f), \quad B(f,\phi) = H(f) \cdot \delta(\phi) \end{array}\right\} \quad (4.22)$$

In essence, the above equation only indicates once again that the time-varying channel causes expansion in the frequency domain, and $\delta(\phi)$ means that the time-invariant channel will not expand in the frequency domain.

Although the deterministic time-varying channel expands the signal in both frequency domain and time domain, it is a deterministic effect. If these effects can be measured accurately, in principle, the original signal can still be recovered by some operation in the receiver. However, the time-varying effect of randomness is not so. It leads to the loss of information in the transmission process and the reduction of processor gain.

As a time-varying filter, acoustic channel performs many extremely complex transformations on the input signal. In principle, these transformations can be divided into two categories.

(1) Deterministic transformation. It can be multipath effect, layered medium refraction effect, layered seabed reflection effect, convergence area and dispersion effect of sound channel, and so on. If the sonar processing system can measure or perceive these transformations in real time, they will not reduce the information of the target signal. As long as the complexity of the processor is increased, the information carried by the signal can still be recovered. For example, the adaptive system may perceive the multipath effect of the channel in real time, so as to make full use of the energy reached by various ways.

(2) Randomness transformation. Random interface, random inhomogeneity of medium, such as scattering effects caused by temperature microstructure, turbulence, internal waves, plankton and bubbles. They can only be described by statistical methods, and the sonar processor cannot measure and perceive the transformation effect they produce in real time. These transformations cannot be compensated in the processor, so they cause the reduction of target information and reduce the system performance.

The channel with deterministic transformation is called coherent channel, and the channel with random transformation is called incoherent channel. In fact, ocean acoustic channel is a partially coherent channel.

4.4 System Function of Random Time-Varying Channel

For random time-varying channels, the four system functions defined in the previous section should be regarded as random functions and can only give statistical description. The simplest and powerful tool is the correlation function. Four correlation functions can be used as the system function of random time-varying channel, namely:

$$R_h(\tau, \tau'; t, t') = \langle h(\tau, t)h^*(\tau', t') \rangle$$
$$R_H(f, f'; t, t') = \langle H(f, t)H^*(f', t') \rangle$$
$$R_B(f, f'; \phi, \phi') = \langle B(f, \phi)B^*(f', \phi') \rangle$$
$$R_s(\tau, \tau'; \phi, \phi') = \langle s(\tau, \phi)s^*(\tau', \phi') \rangle$$

These four correlation functions are also Fourier transform pairs. For example:

$$R_H(f, f'; t, t') = \iint R_h(\tau, \tau';\ t, t')e^{j2\pi(\tau'f'-\tau f)}\mathrm{d}\tau\mathrm{d}\tau' \tag{4.23}$$

$$R_s(\tau, \tau'; \phi, \phi') = \iint R_h(\tau, \tau';\ t, t')e^{j2\pi(t'\phi'-t\phi)}\mathrm{d}t\mathrm{d}t' \tag{4.24}$$

Fig. 4.8 Relationship between system functions of random time-varying channel

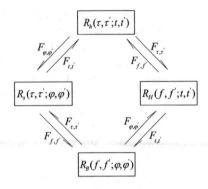

The rest of the Fourier transformation pairs can be written by the readers themselves. The mutual transformation relationship of the four system functions is shown in Fig. 4.8.

Given the autocorrelation function of the input signal, the correlation function of the output of the random time-varying channel can also be expressed by the system function of the channel. According to Eq. (4.21):

$$R_w(t, t') = \langle w(t)w^*(t') \rangle$$

That is,

$$
\begin{aligned}
R_w(t, t') &= \iint \langle z(t-\tau)z^*(t'-\tau') \rangle R_h(\tau, \tau'; t, t')d\tau d\tau' \\
&= \iint R_z(t-\tau, t'-\tau')R_h(\tau, \tau'; t, t')d\tau d\tau'
\end{aligned}
\tag{4.25}
$$

Or,

$$R_w(t, t') = \iint R_z(f, f')R_H(f, f'; \ t, t')e^{j2\pi(ft-f't')}df df' \tag{4.26}$$

where

$$R_z(t-\tau, t'-\tau') = \langle z(t-\tau)z^*(t'-\tau') \rangle$$
$$R_Z(f, f') = \langle Z(f)Z^*(f') \rangle$$

$R_z(t-\tau, t'-\tau')$ is the autocorrelation function of the signal, and RFF is the power spectrum of the signal.

Or use the fourth equation of Eq. (4.21) and take Fourier transformation on both sides of the equation to obtain:

$$R_W(f, f') = \langle W(f) \cdot W^*(f') \rangle$$

$$= \iint \langle Z(f - \varphi)Z(f' - \varphi')\rangle R_B(f, f'; \varphi, \varphi')\mathrm{d}\varphi\mathrm{d}\varphi'$$

$$= \iint R_Z(f - \varphi, f' - \varphi')R_B(f, f'; \varphi, \varphi')\mathrm{d}\varphi\mathrm{d}\varphi' \tag{4.27}$$

where $R_W(f, f')$ is the power spectrum output by the random filter.

$$R_Z(f - \varphi, f' - \varphi') = \langle Z(f - \varphi)Z^*(f' - \varphi')\rangle$$

$R_w(t, t')$ and $R_W(f, f')$ are Fourier transforms of each other.

4.5 Generalized Stationary Channel (WSS Channel), Uncorrelated Scattering Channel (US Channel) [4]

If the channel system function is only related to the time difference and independent of the absolute time, the channel is said to be generalized stationary in the time domain, then:

$$R_h(\tau, \tau'; t, t') = R_h(\tau, \tau'; \Delta t) \tag{4.28}$$

$$R_H(f, f'; t, t') = R_H(f, f'; \Delta t) \tag{4.29}$$

The Fourier transformation relationship according to Eq. (4.24) is:

$$R_s(\tau, \tau'; \phi, \phi') = \iint R_h(\tau, \tau'; t, t')e^{j2\pi(\phi't' - \phi t)}\mathrm{d}t\mathrm{d}t' \tag{4.30}$$

Let $t' = t - \Delta t$, then the above formula becomes:

$$R_s(\tau, \tau'; \phi, \phi') = \iint R_h(\tau, \tau'; \Delta t)e^{j2\pi(\phi' - \phi)t - j2\pi\phi'\cdot\Delta t}\mathrm{d}t\mathrm{d}(\Delta t)$$

$$= R_s(\tau, \tau'; \phi')\delta(\phi' - \phi) \tag{4.31}$$

where

$$R_s(\tau, \tau'; \phi) = \int R_h(\tau, \tau'; \Delta t)e^{-j2\pi\varphi\cdot\Delta t}\mathrm{d}(\Delta t) \tag{4.32}$$

Similarly, it can be proved by Eq. (4.23):

$$R_B(f, f'; \varphi, \varphi') = R_B(f, f'; \varphi)\delta(\varphi - \varphi') \tag{4.33}$$

where

$$R_B(f, f'; \varphi) = \int R_H(f, f'; \Delta t)e^{-j2\pi\varphi \cdot \Delta t} d(\Delta t) \qquad (4.34)$$

As can be seen from Eqs. (4.31) and (4.33), for the generalized time stationary channel, when $\phi \neq \phi'$, R_s and R_B are zero, which means that the scattering components with different Doppler frequencies generated by the sound wave with the frequency of generalized time stationary channel are irrelevant. In short, the generalized time stationary channel must be Doppler uncorrelated scattering channel. This channel is called WSS (wide sense station) in the literature.

In the underwater acoustic field, the normalized sea bottom reverberation and volume reverberation typically meet the WSS conditions. When the sea state is stable, the sea surface reverberation is also generalized time stable, so the reverberation scattering components of different Doppler frequencies are irrelevant.

The other is the generalized stationary channel in frequency domain and uncorrelated scattering channel in time delay domain, which is called us channel for short. This channel is composed of numerous independent scatterers. Because these scattering signals are independent of each other, their superposition will not produce sharp frequency characteristics. Therefore, the total scattering energy has no significant frequency selectivity, that is, it is generalized stationary in the frequency domain.

Readers can follow the discussion on WSS channel and prove by themselves that the system function of US channel can be expressed as:

$$R_h(\tau, \tau'; t, t') = R_h(\tau; t, t')\delta(\tau - \tau') \qquad (4.35)$$

$$R_H(f, f'; t, t') = R_H(\Delta f; t, t') \qquad (4.36)$$

$$R_s(\tau, \tau'; \phi, \phi') = R_s(\tau; \phi, \phi')\delta(\tau - \tau') \qquad (4.37)$$

$$R_B(f, f'; \phi, \phi') = R_B(\Delta f; \phi, \phi') \qquad (4.38)$$

The system functions of the US channel listed in the above formulas are only related to Δf, and the correlation functions in the delay domain have the form of function δ, which means that the channel is generalized stationary in the frequency domain, and the scattering with different delays is uncorrelated.

4.6 Generalized Stationary Uncorrelated Scattering Channel (WSSUS Channel)

When the channel satisfies both generalized time stationary and generalized frequency stationary conditions, this kind of channel is called generalized stationary uncorrelated scattering channel, which is called WSSUS channel for short.

WSSUS channel is a special case of random channel. Its system function has the following forms:

$$R_h(\tau, \tau'; t, t') = R_h(\tau, \Delta t)\delta(\tau - \tau') \tag{4.39}$$

$$R_B(f, f'; \phi, \phi') = R_B(\Delta f, \phi)\delta(\phi - \phi') \tag{4.40}$$

$$R_s(\tau, \tau'; \phi, \phi') = R_s(\tau, \phi)\delta(\tau - \tau')\delta(\phi - \phi') \tag{4.41}$$

$$R_H(f, f'; t, t') = R_H(\Delta f, \Delta t) \tag{4.42}$$

WSSUS channel has only two random variables. All four system functions are only related to Δt and Δf and independent of the absolute value of t and f, which is the direct conclusion of the generalized stationarity of the channel in time domain and frequency domain. The system function is broadened in both delay domain and Doppler domain, but the scattering components of different delay and Doppler frequencies are irrelevant.

$R_s(\tau, \phi)$ is called the scattering function of the channel and $R_H(\Delta f, \Delta t)$ is called the coherence function. These two system functions are very useful in sonar technology. Their physical significance will be explained in the following sections. $R_h(\tau, \Delta t)$ is the cross-correlation function in the delay domain, $R_B(\Delta f, \varphi)$ is the cross-power spectral density function in the Doppler domain. When $\Delta t = 0$ or $\Delta f = 0$, they are called pulse power response $R_h(\tau, 0)$ and $R_B(0, \phi)$ Doppler power spectrum respectively, and the former is the autocorrelation function of impulse response.

Under the condition of WSSUS, the four system functions are Fourier transform each other, and their transformation relationship is shown in Fig. 4.9. A pair of Fourier transformation relations is as follows:

$$R_s(\tau, \phi) = \iint R_H(\Delta f, \ \Delta t)e^{j2\pi(\Delta f \tau - \phi \Delta t)}\mathrm{d}(\Delta f)\mathrm{d}(\Delta t) \tag{4.43}$$

Fig. 4.9 Fourier transform relationship of WSSUS channel system function

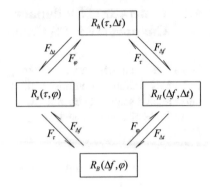

4.7   Scattering Function

In this section, we first investigate the signal passing through the WSSUS channel. Let the sound source radiation signal be $z(t)$ and the received signal be $w(t)$. Since the channel is random, the correlation function $w(t)$ is used to describe the characteristics of the received signal. According to Eq. (4.25), the correlation function $w(t)$ is:

$$R_w(t, t') = \iint \langle z(t - \tau)z^*(t' - \tau') \rangle R_h(\tau, \tau'; t, t') \mathrm{d}\tau \mathrm{d}\tau' \qquad (4.44)$$

Under the condition of WSSUS, replace Eq. (4.39) into Eq. (4.44), note that $t' = t + \Delta t$, and we obtain:

$$
\begin{aligned}
R_w(\Delta t) &= \iint \langle z(t - \tau)z^*(t' - \tau') \rangle R_h(\tau, \Delta t) \mathrm{d}\tau \mathrm{d}\tau' \\
&= \int R_z(\Delta t - \tau) R_h(\tau, \Delta t) \mathrm{d}\tau
\end{aligned}
\qquad (4.45)
$$

Rewrite the convolution on the right of Eq. (4.45) into Fourier integral. Note that the correlation function and power spectrum are Fourier transform each other, and $R_h(\tau, \Delta t)$ and $R_H(\Delta f, \Delta t)$ are Fourier transform each other, that is:

$$R_w(\Delta t) = \int R_Z(\Delta f, \Delta t) R_H(\Delta f, \Delta t) e^{j2\pi \Delta f \Delta t} \mathrm{d}(\Delta f) \qquad (4.46)$$

where $R_Z(\Delta f, \Delta t)$ refers to cross power spectrum of signal $z(t)$.

Then, what is the output response of the matched filter in the WSSUS channel if the processor adopts the matched filter?

Matched filter and copy correlator are equivalent. For the set time delay τ_0 and the Doppler frequency offset ϕ_0, the impulse response function of the matched filter is:

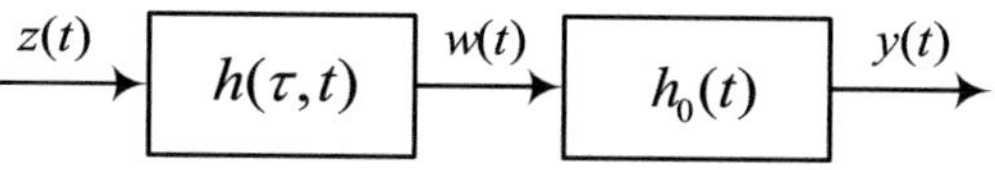

Fig. 4.10 Matched filter in time-varying channel

$$z(t) \rightarrow \boxed{h(\tau, t)} \xrightarrow{w(t)} \boxed{h_o(t)} \xrightarrow{y(t)}$$

$$h_0(t) = z^*(\tau_0 - t)e^{j2\pi\phi_0 t}$$

As can be seen from Fig. 4.10 in the time-varying channel, the output of the matched filter $y(t)$ shall be the convolution between $w(t)$ and $h_0(\tau)$:

$$y(t) = \int w(\beta)z^*(\beta + \tau_0 - t)e^{j2\pi\phi_0(t-\beta)}d\beta$$

Substituting Eq. (4.21) into the above equation, we obtain:

$$y(t) = \iiint s(\tau, \phi)z(\beta - \tau)z^*(\beta + \tau_0 - t)e^{j2\pi\phi_0(t-\beta)+j2\pi\phi\beta}d\beta d\tau d\phi$$

The correlation function of $y(t)$ should be:

$$\langle y(t)y^*(t)\rangle = \iiint \iiint R_s(\tau, \varphi)\delta(\tau - \tau')\delta(\varphi - \varphi')z(\beta - \tau)$$
$$\times z^*(\beta' - \tau')z^*(\beta + \tau_0 - t)z(\beta' + \tau_0 - t)$$
$$\times e^{-j2\pi\varphi_0(\beta-\beta')+j2\pi(\varphi\beta-\varphi'\beta')}d\beta d\beta'd\tau d\tau'd\varphi d\varphi'$$

According to Eq. (3.19), the ambiguity function of signal $z(t)$ is:

$$\chi(\tau, \phi) = \int z(\beta)z^*(\beta + \tau)e^{-j2\pi\phi\beta}d\beta$$

Substituting it into the above equation, we obtain:

$$\langle y(t)y^*(t)\rangle = \iint R_s(\tau, \varphi)|\chi[\tau_0 - \tau + t, \varphi_0 - \varphi]|^2 d\tau d\varphi$$

When $t = 0$, that is, when the filter works at matching points τ_0 and ϕ_0, there is:

$$E(\tau_0, \varphi_0) \triangleq \langle y(t) \cdot y^*(t)\rangle_{\tau_0, \varphi_0} = \iint R_s(\tau, \varphi)|\chi(\tau_0 - \tau, \varphi_0 - \varphi)|^2 d\tau d\varphi \quad (4.47)$$

That is, the output correlation function of the matched filter or copy correlator in the WSSUS channel is the two-dimensional convolution of the ambiguity function of the sound source radiation signal and the channel scattering function. Equation (4.47) can be abbreviated as:

$$E(\tau_0, \phi_0) = R_s(\tau_0, \phi_0) * |\chi(\tau_0, \phi_0)|^2$$

$$= |\chi(\tau_0, \phi_0)|^2 * R_s(\tau_0, \phi_0) \tag{4.48}$$

Equation (4.48) indicates that as long as the pushpin function signal is used, the matched filter can measure the scattering function of the channel. For the pushpin function, its ambiguity function is:

$$|\chi(\tau, \phi)| = \delta(\tau, \phi)$$

Substituting it into Eq. (4.48), it can be seen that the output correlation function of the matched filter is:

$$E(\tau_0, \phi_0) = R_s(\tau_0, \phi_0)$$

The scattering function $R_s(\tau, \varphi)$ of the channel will not be an δ function. Therefore, even if the signal has the ambiguity function of the δ function without measurement ambiguity, the randomness of the channel will reduce the measurement performance of the matched filter and cause additional "ambiguity" to the measurement. The scattering function gives the limit of measurement accuracy caused by the channel, which expresses the inherent resolution limit of the channel in the range (delay) and Doppler (target velocity) planes. If there is a uniform point target (τ_0, ϕ_0), it is no longer regarded as a point in the range Doppler plane in the WSSUS channel, but a "cloud" around (τ_0, ϕ_0), and $R_s(\tau, \phi)$ defines the energy density distribution of the "cloud".

$E(\tau_0, \phi_0)$, also known as the joint ambiguity function of signal and channel, represents the ability of signal and its copy correlator to measure target speed and distance in the channel.

In order to further explain the physical meaning of the scattering function, especially for the convenience of measurement at sea, we define:

$$L = \iint R_s(\tau, \varphi) d\tau d\varphi \Big/ \int R_s(0, \varphi) d\varphi \tag{4.49}$$

$$B = \iint R_s(\tau, \varphi) d\tau d\varphi \Big/ \int R_s(\tau, 0) d\tau \tag{4.50}$$

Let

$$R_s(\tau) = \int R_s(\tau, \phi) d\phi \tag{4.51}$$

$$R_s(\tau) = \int R_s(\tau, \phi) d\phi \tag{4.52}$$

Substituting the above two equations into Eqs. (4.49) and (4.50), we obtain:

$$L = \int R_s(\tau)d\tau / R_s(\tau)|_{\tau=0} \tag{4.53}$$

$$B = \int R_s(\phi)d\phi / R_s(\phi)|_{\phi=0} \tag{4.54}$$

L is the time spread length or time ambiguity length of WSSUS channel, and B is the Doppler spread width or Doppler ambiguity width of WSSUS channel. The product of LB is used to characterize the joint spread of channel to delay and Doppler, or the joint ambiguity of channel. The channel expansion degree is often expressed by the width where $R_s(\tau)$ and $R_s(\phi)$ decrease by 6 dB near the main peak.

RST is called the time spread function of WSSUS channel and rsth is called the Doppler spread function of channel. Their physical meaning is explained below.

As can be seen from Fig. 4.9:

$$R_h(\tau, \Delta t) = \int R_s(\tau, \varphi)e^{j2\pi\varphi\Delta t}d\varphi \tag{4.55}$$

$$R_B(\Delta f, \phi) = \int R_s(\tau, \phi)e^{-j2\pi\Delta f\tau}d\tau \tag{4.56}$$

Take $\Delta t = 0$ in Eq. (4.55) and $\Delta f = 0$ in Eq. (4.56) to obtain:

$$R_h(\tau, 0) = \int R_s(\tau, \phi)d\phi$$

$$R_B(0, \phi) = \int R_s(\tau, \phi)d\tau$$

Note Eqs. (4.51) and (4.52), and the above Equation becomes:

$$R_s(\tau) = \int R_s(\tau, \phi)d\phi = R_h(\tau, 0) = \langle |h(\tau, t)|^2 \rangle_t \tag{4.57}$$

$$R_s(\phi) = \int R_s(\tau, \phi)d\tau = R_B(0, \phi) = \langle |B(f, \phi)|^2 \rangle_f \tag{4.58}$$

The last connected equation in the above two equations is obtained according to the definition of. $\langle |h(\tau, t)|^2 \rangle_t$ is the ensemble average power of the receiver waveform when the sound source emits a δ pulse at time $t - \tau$, and $\langle |B(f, \phi)|^2 \rangle_f$ is the Doppler power spectrum of the receiver waveform when the sound source radiates a continuous sinusoidal sound wave with a single frequency of f. The 6 dB widths of the two are L and B respectively, so the joint ambiguity of the channel can be measured as long as explosion pulse and single frequency continuous sine wave are used. Using the signal of δ function ambiguity function can accurately measure the scattering function of the channel, but the measurement means are complex and the above method is relatively simple. Currently, the scattering function is approximately as follows:

$$R_s(\tau, \phi) \approx R_s(\tau) R_s(\phi) \tag{4.59}$$

Scattering function is of great significance in sonar signal processing theory. Its basic characteristics are summarized as follows:

1. Generally speaking, the system function of random time-varying channel must be expressed by four-dimensional correlation function. Only for WSSUS channel, the scattering function can be degenerated into two-dimensional function. Under what conditions can the actual underwater acoustic channel meet the conditions of generalized time and frequency stationarity?

We regard the actual ocean channel as two parallel filters (coherent multipath channel and random scattering channel). Generally speaking, the reverberation scattering channel can meet the WSSUS condition. Other scattering channels can also be considered to meet the WSSUS condition satisfactorily within an appropriate observation time, taking a reasonable observation frequency band.

Ocean acoustic channel is greatly affected by environmental factors, such as sea state, ocean current, internal wave, and hydrological conditions, which have seasonal, weekly and daily changes, and even significant changes in a few hours. Therefore, strictly speaking, the statistical characteristics of various inhomogeneities are not generally stable in time; The scattering process, such as bubble scattering, has frequency selectivity and frequency dependence. Therefore, the condition of generalized frequency stationarity is not strictly established. Within a certain period of time and frequency, it is considered that the WSS can meet the conditions satisfactorily. Or we can divide the scattering process into fast fluctuating component and slow fluctuating component. This view is expressed in mathematical language:

$$R_H(f, f + \Delta f; t, t + \Delta t) = R'_H(f, t) R''_H(\Delta f, \Delta t) \tag{4.60}$$

where R'_H is unstable in time and frequency, but it changes slowly; $R''_H(\Delta f, \Delta t)$ can be regarded as a rapidly fluctuating scattering component, which meets the WSSUS condition. The above view is to regard the random scattered acoustic channel as two filters in series, one is a slowly changing filter and the other is a fast fluctuating generalized statistical stationary filter in time domain and frequency domain. It can also be regarded as the parallel connection of the above two filters, depending on the specific physical conditions.

2. The scattering function of the channel can be measured by using the waveform of the pushpin function. That is, for the sound source radiation signal $z(t)$, its ambiguity function is $\chi(\tau, \phi)$, $\chi(\tau, \phi)$ has a thumbtack shape, the received signal is processed with a copy correlator, and for the WSSUS incoherent scattering component, the output of the correlator is (according to formula (4.48));

$$E(\tau, \phi) = |\chi(\tau, \phi)|^2 * R_s(\tau, \phi)$$

For the coherent component, the output of the copy correlator is $\chi(\tau, \phi) * h(\tau)$ according to Eq. (3.27). Therefore, considering the coherent component and WSSUS component, the output of the copy correlator is:

$$R_{yy}(\tau, \phi) = |\chi(\tau, \phi) * h(\tau)|^2 + |\chi(\tau, \phi)|^2 * R_s(\tau, \phi) \tag{4.61}$$

If the pushpin function waveform is adopted, there is:

$$\chi(\tau, \phi) = \delta(\tau, \phi)$$

Substituting this formula into Eq. (4.61), we obtain:

$$R_{yy}(\tau, \phi) = |h(\tau)|^2 + R_s(\tau, \phi) \tag{4.62}$$

3. The measurement performance of sonar system is limited by the joint ambiguity of signal and channel. Even if an ideal pushpin signal is used, the signal itself has no measurement ambiguity. The measurement accuracy of sonar system is limited by the joint ambiguity of channel scattering function, the time spread function of channel $R_s(\tau)$ limits the measurement accuracy, and the Doppler spread function limits the velocity measurement accuracy. In the reverberation background, the ability of sonar system to detect low-speed targets is limited. Whether in terms of detecting the target or measuring the target parameters, it is not necessary to design the ambiguity function of the signal too sharp, so that it is not necessary to make it narrower than the ambiguity of the channel.

4.8 Coherence Function

Considering WSSUS channel theory, $R_H(\Delta f, \Delta t)$ is called coherence function, also known as time–frequency correlation function. It represents the two-dimensional joint correlation of the channel in time domain and frequency domain. When $\Delta f = 0$, $R_H(\Delta t, 0) = R_H(\Delta t)$ is called time coherence function, while taking $\Delta t = 0$, $R_H(\Delta f)$ is called the frequency coherence function. According to the mutual Fourier transformation relationship of system functions:

$$R_H(\Delta f, \Delta t) = \int R_h(\tau, \Delta t)e^{-j2\pi\Delta f \cdot \tau}d\tau$$

$$= \int\int R_s(\tau, \phi)e^{j2\pi(\phi \cdot \Delta t - \Delta f\tau)}d\phi d\tau \tag{4.63}$$

Equation (4.63) is actually the inverse transformation form of Eq. (4.43). Take $\Delta f = 0$ in Eq. (4.63), it becomes:

$$R_H(0, \Delta t) = R_H(\Delta t) = \iint R_s(\tau, \phi)e^{j2\pi\phi\cdot\Delta t}d\phi \cdot d\tau$$

$$= \int R_s(\phi)e^{j2\pi\phi\cdot\Delta t}d\phi \tag{4.46}$$

It shows that the time coherence function $R_H(\Delta t)$ and the Doppler spread function $R_s(\phi)$ are Fourier transform each other. The Fourier transform of rectangular pulse waveform is its spectrum. The width of rectangular pulse and its spectrum width can be approximately regarded as reciprocal to each other, which is familiar to most readers. Like the relationship between the width of rectangular pulse and its spectrum, the main peak width of channel time coherence function can be approximately estimated as the reciprocal of Doppler frequency spread width, that is, the coherence time length of channel and Doppler ambiguity width are reciprocal to each other.

Similarly, it proves that $R_H(\Delta f)$ and $R_s(\tau)$ are Fourier transform each other. Therefore, the coherent frequency width of the channel can be approximately estimated as the reciprocal of the width of the delay time spread function $R_s(\tau)$.

The coherent time length and coherent bandwidth (i.e. coherent frequency width) of the channel are very important concepts. If WSSUS channel is a random channel, what do the coherence time length and coherence bandwidth mean?

In fact, the transmission function $H(f, t)$ of the time-varying channel is defined as the complex envelope of the signal generated by the harmonic point source at the receiving point. $R_H(0, \Delta t)$ is the autocorrelation function of the complex envelope waveform, and the coherence time length is the time correlation radius. Therefore, the scattering component can still be regarded as coherent within the observation time less than the coherence time length. In other words, the coherence time length of the channel can be regarded as the average fluctuation period of the WSSUS channel, and the amplitude and phase of the received signal will not fluctuate significantly in a time interval less than the coherence time length. For example, the average period of the sea wave is about 10 s, and the Doppler spread width of the sea surface scattered sound wave is about 0.1 Hz. Its reciprocal, that is, the coherent time length of the sea surface scattered component is 10 s. In the time scale of tens of milliseconds or even seconds, the sea surface can be almost regarded as static, and the received wave shape will not have significant amplitude and phase fluctuations in this time scale. Therefore, as long as the integration time of the coherent processor of the sonar system is short enough, the concept of coherent processing is also effective for the scattering component. Similarly, the coherent bandwidth of the channel can be understood as sending out two harmonic acoustic waves with frequency difference Δf at the same time. When Δf is less than the coherent bandwidth, the phase and amplitude fluctuations of the two received waveforms are related; When Δf is greater than the coherent bandwidth, the phase and amplitude fluctuations will not be correlated.

The ocean channel is partially coherent and partially incoherent. According to the model in Fig. 4.3, we obtain:

$$H(f, t) = H_1(f) + H_2(f, t)$$

where $H_1(f)$ represents the transmission function of coherent channel and $H_2(f, t)$ represents the transmission function of incoherent filter. Following the definition of sound field coherence coefficient in Eq. (4.11), we can define the coherence coefficient of channel as:

$$r_t = |H_1(f)|^2 / (|H_1(f)|^2 + \langle |H_2(f, t)|^2 \rangle_t) \tag{4.65}$$

The coherence coefficient is used to describe the coherence of the channel. As can be seen from Eqs. (4.61) and (4.62), for the sonar system using copy correlator, although the coherent component of the channel also produces measurement ambiguity, the two kinds of ambiguity are essentially different, and the ambiguity generated by incoherent channel is determined by random transformation; The time delay measurement ambiguity caused by coherent channel is a deterministic transformation. As long as the sonar system is complicated, this ambiguity can be overcome in principle. The relevant technologies will be discussed in Chap. 5.

4.9 Directivity Function of Array

Underwater acoustic array is used to measure the directional distribution of sound sources in the sound field. The exact term should be: underwater acoustic array is used to estimate the angular spectrum of the sound field. The directivity function of the array represents its ability to measure the target orientation.

There is a uniform linear array with a length of L in the unbounded space, as shown in Fig. 4.11. The harmonic plane sound wave forms an angle θ with the normal of the array and is projected onto the array. Taking the coordinate origin as the reference point, if the received sound pressure at this point is $e^{j2\pi ft}$, the path difference between point x on the linear array and the reference point is $x \cdot \sin\theta$, and the corresponding phase difference is $\frac{2\pi f}{c} \cdot x \sin\theta$, then the sound pressure received by the array element with length dx at point x on the linear array is:

$$e^{-j2\pi(f/c)\cdot\sin\theta\cdot x + j2\pi ft}\,dx$$

Integrating the above formula with respect to x, we obtain the total sound pressure output by the linear array (omitting the time factor $e^{j2\pi ft}$) as follows:

Fig. 4.11 Plane wave projected on a uniform linear array

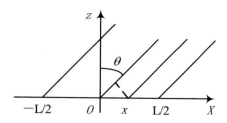

$$p(f, u) = \int\limits_{-L/2}^{L/2} e^{-j2\pi ux} \mathrm{d}x \qquad (4.66)$$

where $u = \frac{f}{c} \cdot \sin\theta$, c is the sound velocity.

Equation (4.66) shows that the total output of the array is related to the incident angle of plane sound wave. When the sound source is directly in front of the array, $\theta = 0$, the maximum output $p(f, 0) = L$ is obtained. Equation (4.66) divided by L is the normalized array directivity function. Except for an unimportant constant coefficient, Eq. (4.66) is the directivity function of uniform linear array, Without losing generality, the integral of the imperative Eq. (4.66) is the directivity function $e(f, \theta)$, so there is:

$$e(f, u) = \int\limits_{-L/2}^{L/2} e^{-j2\pi ux} \mathrm{d}x = \int\limits_{-\infty}^{+\infty} Q(x)e^{-j2\pi ux} \mathrm{d}x \qquad (4.67)$$

where $Q(x)$ is called the aperture function of the array, and for a uniform linear array:

$$Q(x) = \begin{cases} 1 & -L/2 \le x \le L/2 \\ 0 & others \end{cases}$$

Equation (4.67) is Fourier transform, so the directivity function and aperture function of linear array are Fourier transform pairs. Compared with Fourier transform in time domain and frequency domain, the time waveform of the signal corresponds to the aperture function of the array, so the function $e(f, u)$ can also be the angular wave vector spectrum estimation of the array to the point source sound field, and the variable u is called the spatial angular frequency. When the uniform linear array is infinite, $Q(x) = 1$, and Eq. (4.67) becomes δ function, which shows that there is no error in the estimation of angular wave vector spectrum. If the sound field is not uniform, for example, there are multiple distant sound sources in the sound field, or there is multi-path arrival. Assuming that the sound pressure distribution of the sound field along the linear array is $H(f, x)$, the output of the array in the sound field can be obtained from the modified Eq. (4.67):

$$D(f, u) = \int\limits_{-\infty}^{+\infty} Q(x) \cdot H(f, x)e^{-j2\pi ux} \mathrm{d}x \qquad (4.68)$$

The above equation is the response of the array in any sound field, and $D(f, u)$ gives the angular wave vector spectrum estimation of the array to the sound field.

The previous discussion can be easily extended to the case of two-dimensional array. Let the plane array be placed in the vertical plane, x is the horizontal coordinate, y is the vertical coordinate, and its aperture function is $Q(x, y)$, then its directivity function is:

$$e(f, u, v) = \iint Q(x, y)e^{-j2\pi(ux+vy)}dxdy \qquad (4.69)$$

where $u = \frac{f}{c} \cdot \sin\theta_H$, $v = \frac{f}{c} \cdot \sin\theta_V$, θ_H is the angle between the projection of the target diameter vector in the horizontal plane and the normal of the array; θ_V is the angle between the target diameter vector and the normal of the array in the lead plane.

Equation (4.69) shows that the directivity function of the planar array is the Fourier transform of its aperture function.

If the sound pressure distribution function on the receiving surface of the planar array is $H(f, x, y)$, the output of the array with the directivity function $e(f, u, v)$ in the sound field is:

$$D(f, u, v) = \iint Q(x, y)H(f, x, y)e^{-j2\pi(ux+vy)}dxdy \qquad (4.70)$$

The output $D(f, u, v)$ of the array is essentially the estimation of the angular wave vector spectrum of the sound field by the array.

Equations (4.68) and (4.70) are similar. From the perspective of acoustic theory, the above discussion is not rigorous and fine, and a more in-depth discussion is beyond the purpose of this book.

4.10 Random Space Variant Channel

The acoustic channel is not only time-varying but also space-varying. This section will systematically introduce the basic concept of space-varying channel. Therefore, regardless of the time-varying of the channel, it is assumed that the channel is only space-varying.

Let the radial vector of the sound source position be s, the receiving point be r, and the frequency of the harmonic sound wave emitted by the sound source be f. Based on the time-varying channel, the transmission function of space-varying channel can be defined as $H(f, s, r)$ and impulse response function $h(\tau, s, r)$. We can only investigate the case where the sound source is fixed. Therefore, the position diameter vector s of the sound source can be omitted in the following description.

The simplest and most useful case is one-dimensional channel. Assuming that the channel characteristics are only described by the sound field at the horizontal coordinate x, it is called one-dimensional space-varying channel. Following the time-varying channel, four system functions can be used to describe the space-varying

Fig. 4.12 Four system functions of one-dimensional space variant channel

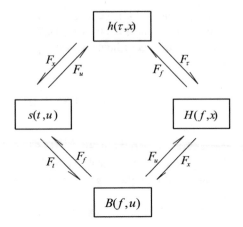

channel, only one of which is independent. The four system functions are Fourier transform pairs of each other, and their relationship is shown in Fig. 4.12

If the far-field harmonic point source is used to measure the characteristics of the space variant channel, and the sound pressure at the receiving point is $H(f, x)e^{j2\pi ft}$, $H(f, x)$ can be defined as the transmission function of the space variant channel; If the sound source radiates δ pulse and the sound pressure at the receiving point is $h(\tau, x)$, $h(\pi, x)$ is called the impulse response function of space variant channel.

As can be seen from Fig. 4.12, the other two system functions are:

$$B(f, u) = \int H(f, x)e^{-j2\pi ux}dx \qquad (4.71)$$

$$s(\tau, u) = \int h(\tau, x)e^{-j2\pi ux}dx \qquad (4.72)$$

Comparing Eqs. (4.71) with (4.68), it shows that the dual frequency function $B(f, u)$ of space variant channel is the output of the envisaged infinite uniform linear array along the x axis in the harmonic point source sound field $H(f, x)$. At this time, $Q(x) \equiv 1$, and $u = \frac{f}{c} \cdot \sin \theta_H$, θ_H is the angle between the horizontal projection of the target radial vector and the normal of the array. Therefore, the dual frequency function is the angular wave vector spectrum estimation of an infinite uniform linear array for an arbitrary harmonic sound field. The estimation of angular wave vector spectrum of sound field by ideal infinite linear array is error free. If the channel is ideal, $B(f, u)$ should be $\delta(u - u_0)$ for plane wave sound field with wave vector u_0. If the space variant channel is coherent and multi-path, $B(f, u)$ is a deterministic function, which produces the broadening of the angular wave vector spectrum for the point source, but this is a deterministic transformation. As long as the transformation can be measured accurately, it only needs to complicate the array and processor, which can overcome the target angular spectrum broadening caused by the channel in theory.

$s(\tau, u)$ describes the spatial angular frequency distribution of the response of space variant channel to the point source of radiation pulse δ at the output. It describes the expansion of channel in spatial frequency domain, so it is called space-varying channel expansion function.

What is the DF performance of an array with an aperture function of $Q(x)$, that is, an array with a directivity function of $e(f, u)$, in space-varying channels? The estimation of the angular wave vector spectrum of the sound field by the array can be determined by Eq. (4.70). This formula can also be expressed in convolution form. Just note that $Q(x)$ and $e(f, u)$ and are Fourier transform each other, and $H(f, x)$ and $B(f, u)$ are Fourier transform each other, then there is:

$$D(f, u) = e(f, u) * B(f, u) \tag{4.73}$$

Equation (4.73) is the one-dimensional convolution of u, and the azimuth estimation of point target by linear array is the one-dimensional convolution of directivity function and dual frequency function.

It can be seen from the above analysis that the azimuth estimation of the point target by the array in the space-varying channel is represented by filter banks, as shown in Fig. 4.13.

The sound field in the ocean usually changes slowly along the longitudinal direction (the horizontal direction from the sound source to the receiving point). Therefore, in a more general case, only two-dimensional channels need to be investigated, as shown in Fig. 4.14. For two-dimensional space variant channels, there are four system functions:

x, y are the horizontal and vertical coordinates on the channel cross section respectively. $u = \frac{f}{c} \cdot \sin \theta_H$, $v = \frac{f}{c} \cdot \sin \theta_V$. θ_H represents the angle between the projection

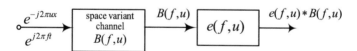

Fig. 4.13 Azimuth estimation of point source in space variant channel by array

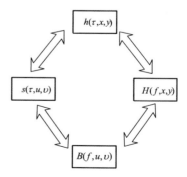

Fig. 4.14 System function of two-dimensional space variant channel

of the sound propagation direction in the horizontal plane and the normal of the array, and θ_V represents the angle between the sound propagation direction and the horizontal plane.

The four system functions are two-dimensional Fourier transforms, for example:

$$B(f, u, v) = \iint H(f, x, y)e^{-j2\pi(ux+vy)}dxdy \tag{4.74}$$

$$H(f, x, y) = \frac{1}{4\pi^2} \iint B(f, u, v)e^{j2\pi(ux+vy)}dudv \tag{4.75}$$

$$u = \frac{f}{c} \cdot \sin\theta_H, v = \frac{f}{c} \cdot \sin\theta_V$$

$H(f, x, y)$ is the sound pressure distribution of each point on the receiving surface, and its two-dimensional Fourier transform $B(f, u, v)$ is the wave vector spectrum of the point source in the space variant channel. The integration of Eq. (4.75) shows that the total sound pressure at the receiving point can be regarded as the superposition of all components scattered along various directions, so it has different wave vectors.

For random space-varying channels, four correlation functions should be used to represent the system function of the channel. The generalized stationary space variant random channel in spatial domain and frequency domain is called WSSUS space variant channel. For WSSUS channel, the four system functions are:

$$\left.\begin{aligned}
R_H(f, f'; x, y, x', y') &= R_H(\Delta f, \Delta x, \Delta y) \\
R_h(\tau, \tau'; x, y, x', y') &= R_h(\tau, \Delta x, \Delta y)\delta(\tau' - \tau) \\
R_B(f, f', u, u', v, v') &= R_B(\Delta f, u, v)\delta(u' - u)\delta(v' - v) \\
R_s(\tau, \tau'; u, u', v, v') &= R_s(\tau, u, v)\delta(\tau' - \tau)\delta(u' - u)\delta(v' - v)
\end{aligned}\right\} \tag{4.76}$$

They are Fourier transforms of each other, for example:

$$R_B(\Delta f, u, v) = \int R_s(\tau, u, v)e^{-j2\pi\Delta f\tau}d\tau \tag{4.77}$$

Taking f = 0, we can get:

$$R_B(0, u, v) = \int R_s(\tau, u, v)d\tau$$

Let

$$R_s(u, v) = \int R_s(\tau, u, v)d\tau = R_B(0, u, v)$$
$$= \langle B^2(\Delta f, u, v)\rangle\big|_{\Delta f=0} \tag{4.78}$$

The last equation of Eq. (4.78) is established according to the definition of R_B. It should be noted that due to the nature of frequency stationarity, the final expression of Eq. (4.78) is independent of f, but only related to frequency difference Δf. $R_s(u, v)$ is called the normalized angular wave vector power spectrum of harmonic point source in the channel, and it is also regarded as the ambiguity function of the channel in the wave vector space. A point source is not a point from the receiving surface, but a "cloud", which is caused by the effect of random multipath scattering.

The array with aperture function $Q(x, y)$ is placed in the xy plane to measure the azimuth of the point source, the output of the array can be seen from Fig. 4.13 as follows:

$$E(u, v) = \langle e(f, u, v) * B(f, u, v) \times e * (f, u, v) * B(f, u, v) \rangle_f$$
$$= \langle |e(f, u, v)|^2 \rangle_f * R_s(u, v) \tag{4.79}$$

In the free sound field, the directivity function $e(f, u, v)$ of the array determines the direction finding performance of the array, which is determined by the aperture function $Q(x, y)$ of the array. The direction-finding performance of the array in the WSSUS channel is determined by the joint ambiguity function $E(u, v)$, which depends not only on the directivity function of the array, but also on the ambiguity function of the channel. If $E(u, v) = \delta(u, v)$, it means that the array has no direction finding ambiguity in the channel. However, Eq. (4.79) shows that even if the directivity of the array is ideal, that is, the directivity function is $\delta(u, v)$ and the joint ambiguity function $E(u, v)$ is $R_s(u, v)$, which indicates that there is still DF ambiguity. Theoretically, the azimuth ambiguity of the channel can be measured by a very narrow directional array, but it is impossible to achieve in practice. Like the reciprocal of the coherent time width and Doppler frequency spread width of the channel in random time-varying channel, the spatial correlation scale of the measurement channel can be used to replace the azimuth ambiguity of the measurement channel in random space-varying channel. Readers can prove by themselves that if the transverse horizontal correlation radius of the channel is L_H and the vertical correlation radius is L_v, the horizontal azimuth ambiguity angle Ω_H and vertical ambiguity angle Ω_V of the channel are:

$$\Omega_H \approx \arcsin \frac{\lambda}{2\pi L_H}, \quad \Omega_V \approx \arcsin \frac{\lambda}{2\pi L_V} \tag{4.80}$$

The transverse correlation radius of the sound field in the ocean is known to be very large, so the measurement accuracy limit of the horizontal orientation of the target is very high; The vertical correlation radius of the sound field is very small, so the measurement error of the vertical angle θ_V of the plane array in shallow water is very large, so it has no practical significance. Therefore, it is impossible to measure the depth of long-range targets in shallow water by relying on the conventional array.

4.11 Physical Model of Passive Sonar

Passive sonar is used to find the sounding target, determine its azimuth, determine the target parameters, and identify the target type. In terms of signal processing, the basic problem concerned is to estimate the space–time structure of the signal field, that is, to estimate the power spectrum and angular wave vector spectrum of the signal field.

The receiving array of sonar system picks up the space–time information of signal field. The structure of array determines the potential ability of sonar system to process information. The channel distorts and blurs the original information sent by the sound source. Therefore, the detection and measurement ability of the sonar system in the channel is limited by the array structure and the channel structure.

Passive sonar must detect target noise (signal) in marine environmental noise field. Target noise is usually ship radiated noise, and one of the important components of environmental noise is also ship traffic noise. Ship noise consists of two parts: one is continuous spectrum, which is dominated by propeller cavitation noise and hydrodynamic noise; The second is the beat sound of propeller and the natural vibration radiation sound of mechanical motion, which are a series of linear spectral components. The intensity of line spectrum component is high. As long as the ship working condition remains unchanged, the frequency of line spectrum is also stable. The target and observer usually move relatively. Doppler effect leads to the frequency shift of line spectrum, and multi-path interference leads to the change of line spectrum shape. Although the ocean channel brings a lot of complexity, the line spectrum carries the information of target types and target motion parameters, and because it has high intensity, using line spectrum to detect and identify targets is an important field of passive sonar technology. Because the line spectrum is very narrow and stable, it stimulates the development of various high-resolution spectrum analysis methods, such as line spectrum tracking technology, autoregressive spectrum estimation technology, adaptive spectrum estimation technology and so on.

The classical concept of passive sonar is "searchlight" sonar. The receiving array of sonar forms a narrow directional beam. When the main beam "illuminates" the target, the outputs of all elements of the array are superimposed in phase after passing through the beamformer. At this time, the output of the array arrives at the maximum. When the main beam deviates from the target, the array outputs environmental noise with low intensity, Sonar detects the target according to the output size of the array. This section will discuss the modern concept of passive sonar from the perspective of acoustic channel theory.

The noise target is assumed to be a point sound source. Its position diameter vector is s, and the random noise waveform is $z(t, s)$. The receiving array is placed in the xy plane to pick up the spatio-temporal information of the target sound field. We assume that the sound field is time invariant and only space-varying channel. If the position radial vector of the receiving point is $\mathbf{r}$, the receiving waveform at this point shall be the convolution of the sound source radiation waveform and the channel impulse response function, which is:

$$\omega(t, \mathbf{r}) = \int z(t - \tau, s)h(\tau, s, \mathbf{r})\mathrm{d}\tau \tag{4.81}$$

Since the sound source position $\mathbf{s}$ is fixed and there is no need to make statistics on the variable $\mathbf{s}$, the writing $\mathbf{s}$ is omitted in the following analysis. Because the signal and channel radiated by the target are random, the target can only be detected and identified according to the spatio-temporal correlation function of the received sound field. The spatiotemporal correlation function $R_w(t, t'; \mathbf{r}, \mathbf{r}')$ of the received sound field is:

$$\begin{aligned}
R_w(t, t'; \mathbf{r}, \mathbf{r}') &= \langle \omega(t, \mathbf{r})\omega^*(t', \mathbf{r}') \rangle \\
&= \iint \langle z(t - \tau)z^*(t' - \tau') \rangle R_h(\tau, \tau', \mathbf{r}, \mathbf{r}')\mathrm{d}\tau\mathrm{d}\tau'
\end{aligned} \tag{4.82}$$

where $\mathbf{r}, \mathbf{r}'$ are two points separated on the xy plane. Assuming that the noise $z(t)$ radiated by the sound source is generalized stationary in time and the space-varying channel is generalized stationary in space, the correlation function R_w and R_h will only be related to $|\mathbf{r} - \mathbf{r}'|$ and independent of the absolute position of the receiver; It will only be related to the time difference $\tau = t' - t$, not to the absolute time. Therefore, Eq. (4.82) becomes:

$$R_w(\tau, |\mathbf{r} - \mathbf{r}'|) = \int R_z(\tau - \beta)R_h(\beta, \rho)\mathrm{d}\beta \tag{4.83}$$

where $\rho = |\mathbf{r} - \mathbf{r}'|$

Fourier transform the variable t on both sides of Eq. (4.83), then we obtain:

$$R_W(f, \rho) = R_Z(f) \cdot R_H(\Delta f, \rho)$$

where $R_W(f, \rho)$ is the cross power spectrum of two received signals at different points of the received sound field, $R_Z(f)$ is the power spectrum of the target signal, and $R_H(\Delta f, \rho)$ is the coherence function of the space variant WSSUS channel. Carrying out two-dimensional Fourier transform on the spatial coordinates on both sides of Eq. (4.83), we obtain:

$$P(f, u, v) = R_Z(f) \cdot R_B(\Delta f, u, v) \tag{4.84}$$

$P(f, u, v)$ is called the wave vector function of sound field. It can be seen from the previous derivation process:

$$P(f, u, v) = \iiint R_w(\tau; \Delta x, \Delta y)e^{-j2\pi(f\tau + u\Delta x + v\Delta y)}\mathrm{d}\tau\mathrm{d}(\Delta x)\mathrm{d}(\Delta y) \tag{4.85}$$

That is, the wave vector function is the power spectrum of the spatiotemporal correlation function of the received sound field in the frequency domain and the angular wave vector domain.

Equation (4.84) shows that in order to obtain the power spectrum information of the target radiated noise (the shape of line spectrum and continuous spectrum, etc.) and the azimuth information of the target (which is included in $R_B(\Delta f, u, v)$), one of the methods is to estimate the spatio-temporal correlation function $R_w(\tau, \Delta x, \Delta y)$ of the sound field and analyze it in frequency domain and angular wave vector domain.

In most sonar applications, only the horizontal azimuth θ_H of the target needs to be measured, so only one-dimensional space-varying channel needs to be investigated. Let the matrix be a linear array arranged along the x direction, then Eqs. (4.84) and (4.85) can be degenerated into:

$$P(f, u) = R_Z(f)R_B(\Delta f, u)$$

$$P(f, u) = \iint R_w(\tau, \Delta x)e^{-j2\pi(f\tau + u\Delta x)}d\tau d(\Delta x) \tag{4.86}$$

To sum up, the task of passive sonar signal processing is to estimate the wave vector function $P_w(f, u, v)$ of the target noise field in the ambient noise field, so as to estimate the power spectrum of the target signal and its angular wave vector spectrum. This is a spatiotemporal spectrum estimation problem. The resolution and accuracy of spectrum estimation are the key problems. In fact, we can't get the statistical function of ensemble average, we can only do statistical average in time domain and space domain. It can only be estimated according to a short time and space sample. The sonar array cannot be too long and the number of array elements cannot be too many. Therefore, the space sample is particularly short. On the other hand, it needs high angular wave vector spectrum resolution in order to obtain good azimuth resolution, and the line spectrum is very narrow, so it needs high resolution. In short, one of the key technologies of passive sonar is to obtain high-resolution and high-precision spectrum estimation based on short samples. It is especially difficult to meet the above requirements under the condition of low signal to clutter ratio, which stimulates the vigorous development of various nonlinear high-resolution spectrum estimation technologies in modern times.

Equation (4.84) points out that the channel will distort the power spectrum shape of the target signal, because R_B also contains frequency variables, and in fact R_B is also related to the depth of the sound source and the depth of the receiving array. The channel will blur the power spectrum information of the target. It is important to study the influence of marine environment conditions on the characteristics of the line spectrum of ship radiated noise for target recognition technology.

The DF accuracy of passive sonar first depends on the estimation accuracy and resolution of diagonal wave vector spectrum, which is related to the power spectrum shape of target signal and the size and structure of array. However, the insurmountable limitation is the ambiguity of channel, and the distortion of channel is difficult to overcome in engineering implementation. The azimuth measurement accuracy of sonar system cannot exceed Ω_H. Equation (4.80) indicates that the magnitude of Ω_H can be estimated according to the spatial correlation radius of the sound field. A series of marine experiments indicate that the magnitude of Ω_H is less than $0.1°$. Ocean acoustic channel has upper and lower interfaces, and multipath effect makes it impossible to measure meaningful θ_v by conventional methods.

4.12 Response of Linear Array in Random Space-Varying Channel

Linear array is the most commonly used form of array in sonar technology. This section discusses its output signal-to-noise ratio in space variant channel and the influence of channel randomness on array processing gain. Discussing this problem is helpful to understand the basic theory of space variant channel. The linear array can be either placed horizontally or vertically. We investigate the case where the linear array is placed along the horizontal X axis.

If the detected target in the space variant channel is a harmonic point source, the aperture function of the linear array is $Q(f, x)$, and the directivity function $e(f, u)$ of the linear array can be obtained from Eq. (4.67):

$$e(f, u) = \int Q(f, x)e^{-j2\pi ux}dx \qquad (4.87)$$

The output of the array can be obtained from Eq. (4.73):

$$D(f, u) = \sigma_s e(f, u) * B(f, u)$$
$$= \sigma_s \int e(f, u - \beta)B(f, \beta)d\beta \qquad (4.88)$$

In the coherent channel, the output power of the array processor is:

$$|D(f, u)|^2 = \sigma_s^2|e(f, u) * B(f, u)|^2 \qquad (4.89)$$

where σ_s^2 represents the target signal power at the array. Equation (4.89) shows that the response of the array to coherent plane waves is added in amplitude.

The output of incoherent received sound field of the array in random space-varying WSSUS channel can be obtained from Eq. (4.78):

$$\langle|D(f, u)|^2\rangle = |e(f, u)|^2 * R_s(u) \qquad (4.90)$$

Equation (4.90) shows that the response of the array to incoherent point sources is summed by power. If the received background interference is the incoherent sound field of WSSUS and the target signal field is coherent, it can be seen from Eqs. (4.89) and (4.90) that the output signal-to-noise ratio of the array processor is:

$$\frac{S_{\text{coherence}}}{N_{\text{non coherence}}} = \frac{\sigma_s^2 |e(f, u) * B_s(f, u)|_{f_s, u_s}^2}{\sigma_N^2 |e(f, u)|^2 * R_{SN}(u)|_{f_s, u_s}} \tag{4.91}$$

where σ_N^2 represents the average interference intensity at the receiving array. The numerator and denominator on the right of the equation take the values at $f = f_s$, $u = u_s$. f_s is the target sound source frequency and $u_s = \frac{f}{c} \cdot \sin \theta_s$, θ_s is the target horizontal orientation.

If the received signal field is also incoherent, the output signal-to-noise ratio of the array processor is:

$$\frac{S_{\text{non coherence}}}{N_{\text{non coherence}}} = \frac{\sigma_s^2 |e(f, u)|^2 * B_{ss}(u)_{f_s, u_s}}{\sigma_N^2 |e(f, u)|^2 * R_{SN}(u)|_{f_s, u_s}} \tag{4.92}$$

Due to the randomness of the channel, the received signal field is transformed from coherent to incoherent. The transformation of the array processing gain is recorded as F_D, then F_D is:

$$F_D = \frac{S_{\text{non coherence}}/N_{\text{non coherence}}}{S_{\text{coherence}}/N_{\text{non coherence}}} = \frac{|e(f, u)|^2 * R_{ss}(u)|_{f_s, u_s}}{|e(f, u) * B_s(u)|^2_{f_s, u_s}}$$

$$\approx \frac{|e(f, u)|^2 * R_{ss}(u)|_{f_s, u_s}}{|e(f, u)|^2_{f_s, u_s}} \tag{4.93}$$

According to Eq. (4.78), we obtain:

$$R_s(u) = R_B(0, u) \tag{4.94}$$

According to the Fourier transform relationship of the system function of random space-varying channel, we obtain:

$$R_B(0, u) = \int R_H(0, \Delta x) e^{-j2\pi u \Delta x} d(\Delta x) \tag{4.95}$$

For the normalized directivity function, when the beam is aimed at the target, there is:

$$|e(f, u)|^2_{f_s, u_s} = 1 \tag{4.96}$$

Considering Eqs. (4.95) and (4.96), Eq. (4.93) can be written as follows by using convolution theorem:

$$F_D = \int \rho(\Delta x) R_H(f_s, \Delta x) e^{-j2\pi u \Delta x} d(\Delta x) \tag{4.97}$$

where $\rho(\Delta x) = \int Q(x)Q(x+\Delta x)dx$ is the correlation function of normalized aperture function. For a uniform linear array with length L, the aperture function is:

$$Q(x) = \begin{cases} 1 & -L/2 \le x \le L/2 \\ 0 & \text{others} \end{cases} \tag{4.98}$$

$$\rho(\Delta x) = \begin{cases} \frac{1}{L}\left(1 - \left|\frac{\Delta x}{L}\right|\right) & -L \le \Delta x \le L \\ 0 & \text{others} \end{cases} \tag{4.99}$$

Equation (4.97) shows that the influence of the randomness of time invariant space-varying channels on the processing gain of the array can be estimated as long as the aperture function of the array and the spatial coherence function of the channel are given. There are many experimental results on spatial coherence function in the literature library. One form of spatial correlation function of harmonic point source sound field is:

$$R_H(\Delta x) = e^{-\frac{1}{2}\left|\frac{\Delta x}{a}\right|^m} \tag{4.100}$$

where, a represents the spatial correlation radius of the sound field, usually m takes 1 or 2.

Substituting Eqs. (4.99) and (4.100) into Eq. (4.97), we obtain:

$$F_D = \frac{1}{L} \int_{-L}^{L} \left(1 - \left|\frac{\Delta x}{L}\right|\right) e^{-\frac{1}{2}\left|\frac{\Delta x}{a}\right|^m} d(\Delta x) \tag{4.101}$$

Letting $\xi = \Delta x/L$ and substituting it into Eq. (4.101) to obtain, we obtain:

$$F_D = \int_{-1}^{1} (1 - |\xi|) e^{-\frac{1}{2}\left|\frac{L\xi}{a}\right|^n} d\xi$$

The above integral can be obtained through calculation:

$$F_D = \begin{cases} 1 & \frac{L}{a} << 1 \\ \frac{4a}{L} & \frac{L}{a} >> 1, m = 1 \\ \sqrt{2\pi} a/L & \frac{L}{a} >> 1, m = 2 \end{cases}$$

Therefore, he analysis of F_D shows that it is inappropriate for the array to be too long and exceed the correlation radius of the sound field in the random channel, and the gain degradation of the array is too serious. The actual ocean channel is partially coherent. At this time, it is still beneficial to increase the array, but only by fully understanding the spatio-temporal correlation function of the sound field and synthesizing sufficient experimental results can the scale of the array be reasonably selected.

References

1. Middleton D. Introduction to statistical communication theory. New York: McGraw-Hill; 1960.
2. Caruthers JW. Fundamentals of marine acoustics. Elsevier; 1977.
3. Steinberg JC, Birdsall TG. Underwater sound propagation in the straits of Flirida. J.A.S.A. 1966;39:301.
4. Laval R. Sound propagation effects on signal processing. In: N.A.T.O proceeding. Loughborough; 1972.

Chapter 5
Slow Time-Varying Coherent Multipath Channel

5.1 Experimental Results of Scattering Function

From the point of view of communication theory, the ocean is regarded as a time-varying and space-varying filter. If the sound source and receiver are fixed, the channel can be regarded as a time-varying filter.

For the slow time-varying coherent channel, the spread function $s(\tau, \varphi)$ can be used to describe the characteristics of the channel; If it is a fast-changing WSSUS channel, the scattering function $R_s(\tau, \varphi)$ is used to describe the characteristics of the channel. As for the partially coherent channel, the output of the copy correlator can be seen from Eq. (4.61):

$$R_{yy}(\tau, \varphi) = |\chi(\tau, \varphi) * h(\tau)|^2 + |\chi(\tau, \varphi)|^2 * R_s(\tau, \varphi) \tag{5.1}$$

where χ is the ambiguity function of the signal. If the thumbtack signal is used, the response of the copy correlator in the channel is:

$$R_{yy}(\tau, \varphi) = |h(\tau)|^2 + R_s(\tau, \varphi) \tag{5.2}$$

In the Eq. (5.2), the first term on the right is zero Doppler component, which represents the energy of coherent component of the channel, and the second term is incoherent component.

In the past 20 years, many scholars have devoted themselves to the research of acoustic channel characteristics, trying to reveal its physical mechanism and study the corresponding optimized sonar signal processor. A series of experiments conducted by the Saclant anti-submarine research center of the North Atlantic Treaty Organization from 1974 to 1978 serve for this purpose. Brief introduction is as follows.

Two thumbtack signals, CW pulse train and linear frequency modulation pulse train, are used successively. The CW pulse width is 40 ms, the repetition period is

© Harbin Engineering University Press 2022
J. Hui and X. Sheng, *Underwater Acoustic Channel*,
https://doi.org/10.1007/978-981-19-0774-6_5

200 ms, and the total length of the sample is 40 min; The linear frequency modulation pulse width is 0.125–2 s, the frequency modulation bandwidth is 500 Hz, the repetition period is 10 s, and the total length of the sample is 20 min. The Doppler frequency resolution of the two thumbtack signals is very high, which is between 0.002 and 0.004 Hz, so that the Doppler spread of the channel can be measured accurately. The total sample length cannot be too long, otherwise the statistical results will deviate from the generalized stationary condition in time domain due to the slow change of environmental parameters. The hydrophone and transmitter were fixed in the experiment. After amplification and pre-filtering, the received signal is copied and correlated by the computer, and the integration time of the correlator is the total length of the signal. Figures 5.1 and 5.2 present interesting measurements of the scattering function.

The zero Doppler main peak in the figure represents the coherent component of the channel. It may be single group or multi group, single peak or multi peak, which depends on the multi-path structure of the channel and is slowly time-varying; The side lobe with Doppler represents the incoherent component of the channel. Generally speaking, its energy is concentrated near the wave frequency, about 0.1–0.2 Hz, which at least indicates that surge is one of the main reasons for the fluctuation of acoustic signal. The main peak is usually more than 10 dB higher than the side lobe, which means that the energy of the coherent component accounts for more

Fig. 5.1 The scattering function. Water depth: 20 m, distance: 3 km; Frequency: 1.5 Hz, CW pulse train

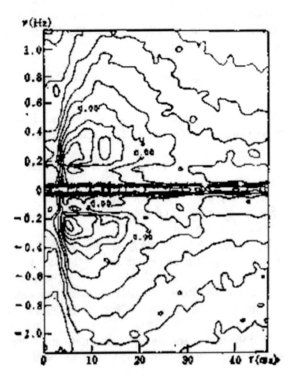

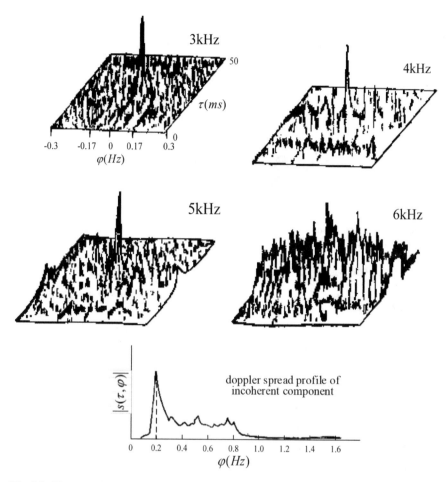

Fig. 5.2 The scattering function (water depth: 60–110 m, distance: 16 km; pulse train frequency modulation, $\Delta f = 500$ Hz, pulse width $= 2$ s)

than 90% of the total energy, which proves that the marine acoustic channel can be regarded as a slow time-varying coherent multipath channel. In most cases, the time delay spread of incoherent components is about 5 ms, and the Doppler spread is about 0.1–0.7 Hz. It can be inferred that the coherence bandwidth is about 200 Hz and the coherence time length is about a few seconds.

The experiment of 6 kHz sound wave shown in Fig. 5.2 is particularly interesting. The non-coherent component of the channel is strong with high sidelobe, and the channel has significant Doppler and delay ambiguity; The coherent component also shows a complex multi-path structure with the characteristics of multiple groups and multiple peaks.

Figures 5.3 and 5.4 show continuous time records of the received signal envelope. The single frequency continuous signal shows a rapid amplitude fluctuation of up

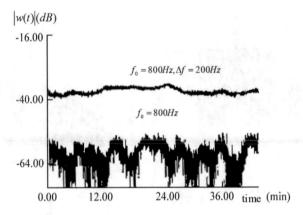

Fig. 5.3 The time recording of received signal envelop (the position of the transmitter and the transducer is fixed)

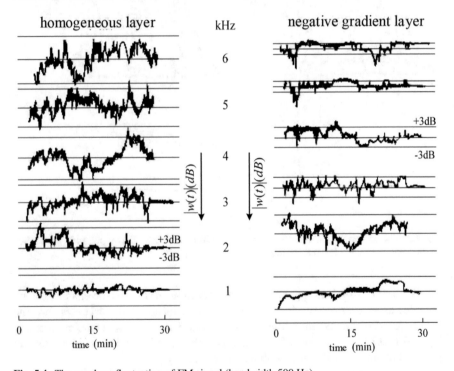

Fig. 5.4 The envelope fluctuation of FM signal (bandwidth 500 Hz)

to 20 dB. It is remarkable that the amplitude fluctuation of white noise with 200 Hz bandwidth is reduced to only 2 dB, which just convincingly shows that the coherent bandwidth of the channel will not exceed 200 Hz, and the fluctuations of sound waves with sufficiently separated frequencies are not related to each other. Therefore, the synthetic amplitude fluctuation of all frequency components in the total passband is very small, and the larger the relative bandwidth is, the smaller the amplitude fluctuation is.

Careful readers can also notice that compared with the upper and lower figures in Fig. 5.3, the low-speed amplitude fluctuation (about 6–8 min average period) in the upper figure is less than 6 dB, while the lower figure is more than 20 dB. The low-speed fluctuation of single frequency continuous wave signal is caused by the slow change of multi-path coherent structure. For a signal with limited bandwidth, its different spectral components have different multi-path interference. After multi-path interference, some spectral components are strengthened and others are weakened. As long as the signal bandwidth is large enough, the fine structure of multi-path interference will be blurred, as a result, the low-speed fluctuation of the limited bandwidth signal envelope is also smoothed. Such case the same for the upper and lower figures: when the signal is strong, the rapid fluctuation is small, when the signal is weak, its rapid fluctuation is large. This phenomenon supports the previous view that the fluctuation characteristics of incoherent components are also related to the interference pattern of coherent components of sound field.

Figure 5.4 shows the envelope fluctuation record of FM signal with 500 Hz bandwidth. It can be seen that the fluctuation of low-frequency signal is very small, which is less than 1 dB, while the amplitude fluctuation at high frequency is in the order of ± 3 dB. Perhaps this fact implies that the incoherent component of low-frequency channel is small, which is consistent with that revealed by the scattering function in Fig. 5.2.

The above experimental results on signal fluctuation have important enlightenment for underwater acoustic Telegraph and coded communication. The communication mode of frequency modulation or pseudo-random noise filling code with hundreds of Hz bandwidth can effectively resist signal fading and help to ensure communication quality.

5.2   Measurement Method of Channel Coherence—Inter Pulse Correlation Method

The characteristics of time-varying channel can be described by any of the four system functions in Fig. 4.8. $R_H(f, f'; t, t')$ directly describes the coherence of the channel. The scattering function of the channel can be measured by the signal of thumbtack function, and then the coherence function $R_H(f, f'; t, t')$ can be obtained by quadratic two-dimensional Fourier transform. However, this is technically complex, and the channel is required to meet the WSSUS condition in principle. The time-varying coherence of the channel can be regarded as the slow time-varying coherence of the channel, so it is easy to give up the time-varying coherence of the channel.

The channel coherence function is defined as:

$$R_H(f, f'; t, t') = \langle H(f, t)H^*(f', t') \rangle$$

The above equation is actually the cross-correlation function of harmonic point source sound field with frequency f and f', which is usually measured by long CW pulse or narrow-band signal. According to Eq. (4.26):

$$R_w(t, t') = \iint R_z(f, f')R_H(f, f'; t, t')e^{j2\pi(ft-f't')}df df' \qquad (5.3)$$

$R_w(t, t')$ is the cross-correlation function of the received waveform and $R_r(f, f')$ is the power spectrum of the sound source radiation signal, and:

$$R_z(f, f') = \langle Z(f)Z^*(f') \rangle$$

The sound source radiation signal is $z(t)$, its spectrum is $Z(f)$, and the received waveform is $\omega(t)$. Assuming that the transmitted signal $z(t)$ is a narrowband signal with a center frequency of f_0, R_H can be proposed as a constant from the integral symbol of Eq. (5.3):

$$R_w(t, t') \approx R_H(f_0, f_0; t, t') \iint R_Z(f, f')e^{j2\pi(ft-f't')}df df'$$
$$\approx R_H(f_0, f_0; t, t')R_z(t, t') \qquad (5.4)$$

If the transmitted signal is a periodically repeated narrow-band deterministic signal $z(t) = z(t + mT)$, T is a repetition period, then:

$$R_z(t, t') = \overline{z(t)z^*(t + mT + \tau)}$$
$$= \overline{z(t)z^*(t + \tau)} = R_z(\tau)$$

Substituting this formula into formula (5.4), we obtain

$$R_w(mT) \approx R_H(f_0, mT)R_z(\tau)|_{\tau=0} = R_H(f_0, mT)R_z(0) \qquad (5.5)$$

The above equation shows that the channel coherence function $R_w(mT)$ can be measured by measuring $R_H(f_0, mT)$.

An example of measuring channel coherence function by inter pulse correlation method is given in Figs. 5.5 and 5.6. The experiment was carried out in shallow water. The transmitting system was set on the ship, the transmitting transducer was fixed on the seabed, the receiving arry was also anchored to the seabed. The transmission

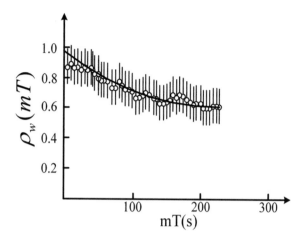

Fig. 5.5 The shallow water coherence function $R_H(f_0, f_0, t, t + mT)$. Transmitted signal: frequency modulation bandwidth is 100 Hz, pulse width is 64 ms, and the cycle is 5S. "Φ" represents the standard deviation of measurement points, which is the statistical result of 22 measurements; $\rho_w(mT) = ae^{-\alpha mT} + b$, $a = 0.37$, $\alpha = 0.011, b = 0.6$

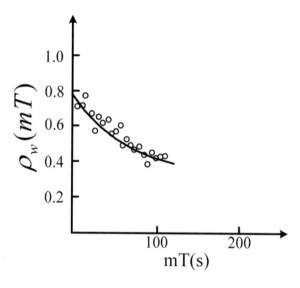

Fig. 5.6 The shallow water channel coherence function. Distance: 12.972 km; "O" is the average value of 6 measurement results; The solid line is the fitting curve: $\rho_w(mT) = ae^{-\alpha mT} + b$, $a = 0.436; \alpha = 0.016; b = 0.36$

period is 5 s, the pulse width of the transmission signal is 640 ms, and the signal form is linear frequency modulation pulse with frequency modulation bandwidth of 100 Hz. The output signal of the receiving array is amplified and filtered and processed in real time with a 4-bit quantized correlator. Each N pulses are divided into a group, $n = 1, 2, \ldots, N$. Using the first received signal as the reference signal of the copy correlator, and performing correlation calculation with the second, third, ..., and Nth received pulse waveforms respectively, $R_{win}(t), i$ can be obtained for $t = (n - 1)T$, i is the group number, and n is the serial number in the group. In order to eliminate the influence of pulse amplitude fluctuation, the normalized correlation coefficient is calculated, namely:

$$\rho_{win} = R_{win}[(n-1)T]/[W_{i2}(0)W_{in}((n-1)T)]$$

In fact, the denominator in the above equation includes the following quantities in the calculation time:

$$W_{i2}(0) = \sqrt{W_{i2}(T)} \quad W_{in}[(n-1)T] = \sqrt{R_{win}[(n-1)T]}$$

Then average the ρ_{wi} of each group according to the group number i:

$$\overline{\rho}_{wn} = \frac{1}{k}\sum_{i=1}^{k}\rho_{win}, n = 1, 2, \ldots, (N-1).$$

Figures 5.7 and 5.8 are the experimental results of two-way channel inter pulse correlation coefficient. The experimental method is the same as the above, but the received signal is the echo of a fixed target.

The experimental results show that the coherence function of the channel decreases exponentially with the increase of time. The experiment reveals that the real ocean channel is a slow time-varying coherent multi-path channel, and its coherence time length is at least a few minutes. It is observed that the channel is partially coherent within dozens of minutes. The time-varying characteristics of the channel coherence function can be illustrated schematically in Fig. 5.9:

1. For the time invariant deterministic coherent channel without interference, we have

$$\rho_w(t) \equiv 1$$

Fig. 5.7 The coherence function of two-way channel. 1—2.965 km, $a = 0.6, \alpha = 0.0104, b = 0.36; 2—56 \times 608 \times 0.3048$ m. $a = 0.36, \alpha = 0.0141, b = 0.30$; Other descriptions are the same as Fig. 5.5

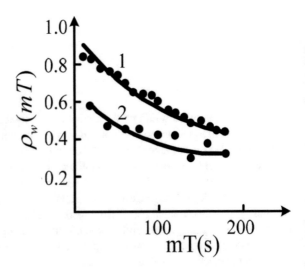

Fig. 5.8 The coherence function of two-way channel. $a = 0.34$, $\alpha = 0.021$, $b = 0.40$; Other descriptions are the same as Fig. 5.5

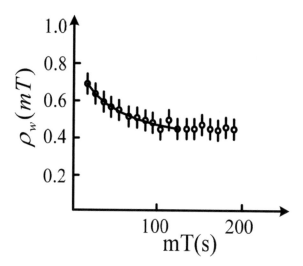

Fig. 5.9 The slowly time-varying coherent channel (T—pulse repetition period)

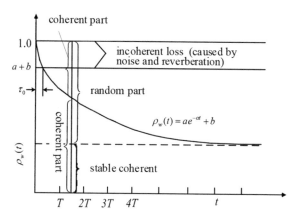

2. For the stable coherent channel with incoherent interference background, due to the interference contained in the received signal, there is a fixed loss of inter pulse correlation, we have: $(a + b) \leq \rho_w(t) < 1$.

3. For slow time-varying coherent channels, the form of inter pulse correlation is:

$$\rho_w(t) = ae^{-\alpha t} + b \tag{5.6}$$

   where a, α, b refer to constants determined by the environmental conditions of the channel and the signal-to-noise ratio.

4. The first point of the correlation coefficient (corresponding to time $t = T$) usually has a large decorrelation rate, which may be due to the correlation loss caused by interference or there may be a fast time-varying process with a coherence time length of several seconds. In this experiment, because $T =$

5 s, it is impossible to make a more deep analysis from this experiment (it can be seen from the previous section of this chapter that there is an incoherent scattering process with a coherent time length of several seconds caused by wave scattering). Experiments show that the inter pulse correlation will tend to form a stable value b. Although the shallow water channel is very complex, even for the more complex two-way channel, it has a stable coherence coefficient. Therefore, the ocean channel is composed of three parts: stable coherent component, slow time-varying coherent component, and fast fluctuating incoherent component.

The inter pulse correlation $\rho_w(T)$ is obtained by using the first received signal waveform as the reference signal of the correlator and the second received waveform for correlation calculation. When T is very small, the two waveforms are nearly the same, and the output of the correlator is equivalent to the signal energy output of the adaptive correlator; When mT is very large, the received signal waveform is significantly different from the first waveform, so the output signal energy of the correlator decreases, which means the received waveform is unknown. Therefore, the size of a is roughly the difference between the output signal of the copy correlator working in the actual ocean channel and the ideal channel, and it is also the energy difference between the adaptive correlator and the copy correlator for signal matching. The adaptive correlator can completely match the channel, and the output signal amplitude should be $(a + b)$, and the signal amplitude output by the copy correlator should be roughly average b. Therefore, the gain difference of signal matching between the two under experimental conditions can be approximately estimated as $20 \log \frac{a+b}{b}$. According to the experimental results, the output signal of the adaptive correlator is 4–8 dB higher than that of the copy correlator. As predicted, the gain difference is related to hydrological conditions.

5.3 Modified Matching of Shallow Water Coherent Multipath Channel with Slow Time Variation

As discussed in Sect. 3.2, the best processor to detect and determine the signal is matched filter or copy correlator with equivalent effect under the background of Gaussian white noise.

In the ideal channel, the received waveform is the same as the transmitted waveform, and the reference signal of the correlator can be copied from the transmitted waveform. Under this condition, the detection effect and measurement performance of the copied correlator appear to be satisfied. In WSSUS channel, the performance of the copy correlator is determined by Eq. (4.48), which has poor matching effect on this part of incoherent energy. For the stable coherent and slowly time-varying coherent components with more than 90% of the total energy, the copy correlator can still be used for coherent processing effectively. In order to match the time-varying coherent part, the reference signal must also be time-varying. The inter pulse correlation method is a simple and effective method when the input signal-to-noise ratio

Fig. 5.10 Comparison of inter pulse correlation and copy correlation. Transmit signal LFM, BT = 50, shallow water channel

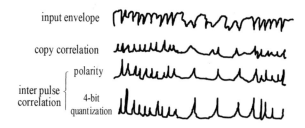

is high. It usually takes the signal received in the previous period as the reference signal of the correlator. If the input signal-to-noise ratio is high, and the repetition period of the signal is significantly shorter than the coherence time length of the slowly changing channel, the inter pulse correlator can often further improve the output signal-to-noise ratio, improve the clarity of the output signal, and significantly reduce the correlation peak and side lobe of the output. The offshore experimental results shown in Fig. 5.10 support the above view.

We will discuss the output SNR of copy correlator and modified correlator in partially coherent channel in the following part. According to the analysis in the previous section, the impulse response function of partial coherent channel is:

$$h(\tau, t) = \sqrt{bE} h(\tau) + \sqrt{aE} \overline{h}(\tau, t) + \sqrt{cE} \tilde{h}(\tau, t) \tag{5.7}$$

where E is the total energy of the received signal, bE, aE, cE represent the signal energy of $h(\tau)$, $\overline{h}(\tau, t)$, and $\tilde{h}(\tau, t)$, which represent the stable coherent part, slowly time-varying coherent part and incoherent part of the channel respectively, and all impulse response functions are normalized, that is:

$$\rho_{hh}(0) = \overline{h(\tau)h^*(\tau)} = 1$$

$$\rho_{\overline{h}\overline{h}}(0, 0) = \overline{\overline{h}(\tau, t)\overline{h}^*(\tau, t)} = 1$$

$$\rho_{\tilde{h}\tilde{h}}(0, 0) = \overline{\tilde{h}(\tau, t)\tilde{h}^*(\tau, t)} = 1 \tag{5.8}$$

According to the results of the marine experiment in the previous section, the energy c of the incoherent component in the actual marine channel is less than 1. Both aE and bE represent the size of coherent components. Their relative size is determined by the specific environmental conditions of the channel. They can be matched as long as the reference signal of the correlator is corrected.

When the transmitted signal is $z(t)$, the received signal will be different from the transmitted signal due to multipath effect and scattering effect. At this time, the received signal is:

$$w(\tau, t) = z(\tau) * h(\tau, t)$$

$$= z(\tau) * \left[\sqrt{bE} h(\tau) + \sqrt{aE} \overline{h}(\tau, t)\right.$$

$$+\sqrt{cE}\tilde{h}(\tau, t)\Big] \tag{5.9}$$

For the correction correlator, the reference signal $x(\tau, t)$ is taken as:

$$x(\tau, t) = w^*(\tau, t)$$

Therefore, the output of the correction correlator is:

$$y(\tau, t) = \overline{w(\tau, t)w^*(\tau, t)}$$

Substituting Eq. (5.9) into the above equation and assuming that the three components are independent of each other, we obtain:

$$y(\tau, t) \approx bE R_{zz}(\tau) * \rho_{hh}(\tau) + aE R_{zz}(\tau) * \rho_{\overline{hh}}\left(\tau, t, t'\right) \tag{5.10}$$

For the correction correlator, when $\tau = 0$, the above formula obtains the maximum value (for inter pulse correlation, take $\tau = T$, T is the pulse repetition period), and the specified signal $z(t)$ is normalized, i.e. $R_{zz}(0) = E_z = 1$, which is obtained from Eq. (5.10):

$$y(0, t) = bE + aE \tag{5.10a}$$

When the interference is white noise, the noise power output by the correction correlator is $N_0 E/2$, and the output signal-to-noise ratio of the correction correlator is:

$$\begin{aligned}(S/N)_{correction} &= \frac{2\big[bE R_{zz}(\tau) * \rho_{hh}(\tau) + aE R_{zz}(\tau) * \rho_{\overline{hh}}(\tau, t, t')\big]^2}{N_0 E} \\ &= \frac{2E(b + a)^2}{N_0}\end{aligned} \tag{5.11}$$

where E is the total energy of received signal and $N_0/2$ is the noise power within 1 Hz bandwidth.

The output signal-to-noise ratio of the copy correlator is:

$$(S/N)_{copy} = \frac{2E_z\Big\{R_{zz}(\tau) * \Big[\sqrt{bE}h(\tau) + \sqrt{aE}h(\tau, t)\Big]\Big\}^2}{N_0} \tag{5.12}$$

Since the coherent channel is assumed to be a multi-path channel, it can be seen from Eq. (3.6):

$$h(\tau) = \sum_{i=1}^{N} A_i \delta(\tau - \tau_{0i}) \tag{5.13}$$

$$\overline{h}(\tau, t) = \sum_{i=1}^{N} A_i'(t) \delta\left[\tau - \tau_{0i}'(t)\right] \tag{5.14}$$

Equations (5.11), (5.12), (5.13) and (5.14) show the processing gain difference between the correction correlator and the copy correlator. In the multipath channel, the output of the copy correlator is multimodal, and the maximum value shall be taken in Eq. (5.12), so G can be estimated as:

$$G = \frac{(b + a)^2}{\left[\max\left\{\sqrt{b}A_i, \sqrt{a}A_i'\right\}\right]^2} \tag{5.15}$$

where $\max\left\{\sqrt{b}A_i, \sqrt{a}A_i'\right\}$ means to take the maximum value in parentheses.

In order to give the quantitative concept of the effect of the correction correlator on channel matching, it is assumed that the multi-path channel is composed of four equal strength channels. Considering Eq. (5.8), that is, the impulse response function of the channel is normalized, it is not difficult to know after simple algebraic operation, $A_i = A_i' = 0.5$. Take the experimental data of Fig. 5.7, $a = 0.36$, $b = 0.30$. According to Eq. (5.15), the channel matching of the modified correlator is 6.85dB higher than that of the copy correlator.

The key to realize the modified correlation matching in time-varying channels is to extract the channel impulse response function in real time. If the channel impulse response function $h(\tau, t)$ is measured in real time, the modified reference signal can be given by Eq. (5.9). When the input SNR is high, the inter pulse correlation method can be used, and when the SNR is low, the adaptive correlation method can be used to obtain the modified correlator reference waveform. Another experiment of correcting correlation matching is briefly introduced here.

Assuming that the transmitted signal is $z(t)$, and there is a high SNR at the receiving point, and the sample is discretized by sampling, the Wiener Hoff equation is expressed as:

$$R_{zz}h = R_{zw} \tag{5.16}$$

Equations (5.16) and (3.31) are the same. R_{zz} is the autocorrelation matrix of the transmitted signal, R is the impulse response vector of the channel, and R_{zw} is the cross-correlation vector of the received signal $\omega(t)$ and the transmitted signal, and:

$$\boldsymbol{R}_{zz}(K) = \begin{bmatrix} z(k)z(K) & z(K)z(K-1)... & z(K)z(K-N) \\ z(K-1)z(K) & z(K-1)z(K-1)... & z(K-1)z(K-N) \\ z(K-N)z(K) & z(K-N)z(K-1)... & z(K-N)z(K-N) \end{bmatrix}$$

(5.17)

$$\boldsymbol{h} = \begin{bmatrix} h_1, h_2, ..., h_{N+1} \end{bmatrix}$$

(5.18)

$$\boldsymbol{R}_{zw}(K) = \begin{bmatrix} \omega(K)z(K) \\ \omega(K)z(K-1) \\ ... \\ \omega(K)z(K-N) \end{bmatrix}$$

(5.19)

The above formulas are estimates of statistics, so there is no time average or ensemble average.

$\boldsymbol{R}_{zz}$ is Eq. (5.16) is already known. $\boldsymbol{R}_{zw}$ is the output of the copy correlator. Only Eq. (5.16) is required to obtain the impulse response vector $\boldsymbol{h}$ of the channel. Equation (5.17) is called Toeplitz type matrix, so Eq. (5.16) is Toeplitz type linear equations, and $\boldsymbol{h}$ can be solved by solving the equations. A fast numerical iterative algorithm for the equations is the L.M.S method. In essence, the adaptive correlator realizes the real-time solution, i.e., Eq. (5.16) by using the L.M.S algorithm. When the weight vector of the adaptive transverse filter reaches stability, it is the minimum mean square error estimation of the channel impulse response vector $\boldsymbol{h}$.

In Ref. [1], the inverse of Toeplitz matrix is directly calculated on the computer to obtain $\boldsymbol{h}$, and an offshore experiment is carried out to test the improvement of modified correlation on channel matching. It is briefly introduced as follows:

The transmitter and receiver are fixed on the shallow sea bed with a distance of 4 N miles. The transmitted signal is 63-bit pulse coding, the symbol width is 10 ms, CW pulse and the center frequency is 1.6 kHz. The received waveform of each symbol can refer to Fig. 5.11 for. According to the output of the copy correlator,

Fig. 5.11 Received waveform in shallow water channel 10 ms CW pulse in 4 N mile

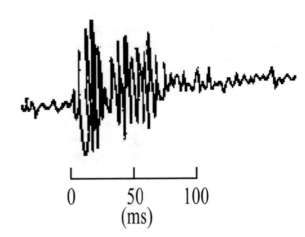

0   50   100
(ms)

the impulse response function of the channel obtained by Eq. (5.16) is shown in Fig. 5.12. Figure 5.13 shows the output waveforms of the copy correlator and the correction correlator drawn by the computer. It can be seen from the figure that the channel matching improvement gain of the modified correlator is about 6 dB, and its sidelobe is significantly smaller than that of the copy correlator. Figure 5.14 shows the channel time-varying coherence function measured simultaneously in the above experiment.

Fig. 5.12 The impulse response function of 4 N mile shallow water channel

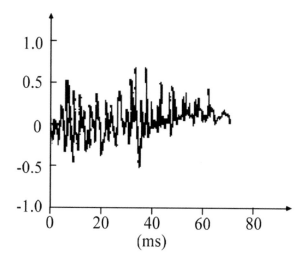

Fig. 5.13 The output waveform of the modified correlation function and the copy correlator

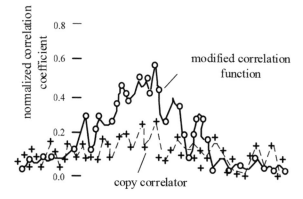

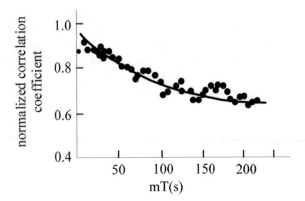

Fig. 5.14 The time-varying channel coherence function in the modified correlation experiment

5.4   Modified Correlation Matching for Deep Sea Coherent Channel

Although the two modified correlation methods for channel matching introduced in the previous section are not practical from the point of view of detection, and the adaptive correlator can detect the target under the condition of low input signal-to-noise ratio, they have the same physical basis for channel coherent matching. It is not redundant to investigate the physical basis of modified correlation and channel matching again for deep-sea channel. The experiments introduced in this section will once again prove that the deep-sea channel is also a slowly time-varying coherent channel. The modified correlator can make full use of the signal energy reached by all kinds of ways, and it has low correlation sidelobe and good measurement and communication capabilities.

In the late 1960s, Columbia University carried out correction related experiments in Bermuda's 5000 m deep sea area. The 32-element receiving array is still placed on the 1200 m deep sound channel axis, and the cable sends the signal to the rock for correction and correlation processing. The transmitting system is placed on a slowly drifting ship, and the transmitting transducer is suspended on the depth of the sound channel axis. The transmitted signal is shown in Fig. 5.15. The leading signal is a CW pulse with a width of 10 ms, followed by a hyperbolic FM signal with a pulse width of 4 s after 6 s, and the center frequency is 300 Hz. Hyperbolic FM signal is a Doppler invariant signal. The ambiguity function of the signal waveform is not sensitive to the target motion. This signal form is used to reduce the influence of the transmitting ship motion. The response waveform of the receiving point to the preamble signal can be approximately regarded as the impulse response function of the channel. Its envelope at 250 N miles is shown in Fig. 5.16.

The hyperbolic FM signal is recorded as $z(t)$ and the corresponding received signal is recorded as $w(t)$. If the system function of the channel does not change within 6 s, the correlation operation between $w(t)$ and $h(\tau)$ can be obtained:

Fig. 5.15 The waveform of transmitted signal

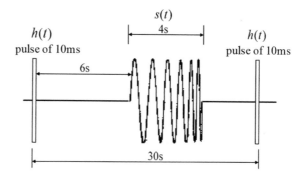

Fig. 5.16 The envelope of impulse response function of 250 N mile channel

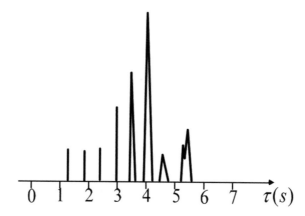

$$y(\tau) = R_{wh}(\tau) = \int_T \omega(t)h(t - \tau)dt \qquad (5.20)$$

where T is the signal width (4 s).

As we all know, $w(t) = z(t) * h(t)$.

Substitute into Eq. (5.20) to obtain:

$$y(\tau) = z(\tau) * h(\tau) * h(-\tau) = z(\tau) * R_{hh}(\tau)$$
$$= R_{hh}(\tau) * z(\tau) \qquad (5.21)$$

where $R_{hh}(t)$ is the autocorrelation function of the impulse response function R_h, that is, the channel power response function R_h. The corrected correlation output can be obtained by performing correlation operation on $y(t)$ and $z(t)$

$$R_{yz}(\tau) = R_{hh}(\tau) * z(\tau) * z(-\tau)$$
$$= R_{hh}(\tau) * R_{zz}(\tau) \qquad (5.22)$$

The above equation is the output of the correction correlator. In comparison, the copy related output is also calculated:

$$R_{wz}(\tau) = R_{zz}(\tau) * h(\tau) \tag{5.23}$$

If the signal bandwidth of $z(t)$ is wide enough, then:

$$R_{zz}(\tau) \approx \delta(\tau)$$

At this time, Eqs. (5.22) and (5.23) are approximately:

$$R_{yz}(\tau) \approx R_{hh}(\tau) \tag{5.24}$$

$$R_{wz}(\tau) \approx h(\tau) \tag{5.25}$$

As we all know, $R_{hh}(0)$ represents the total energy sum of signals arriving in all kinds of ways, and has lower sidelobe than $h(\tau)$, which shows that the modified correlation is better than the copy correlation.

The correction correlation and copy correlation waveforms obtained from offshore experiments prove the above theoretical results, as shown in Figs. 5.17 and 5.18. The experimental results support the view described at the beginning of this section:

1. The modified correlation experimental method shows that under the experimental conditions, the coherence time of deep-sea channel is at least more than 6 s, and the channel is slowly time-varying.
2. The modified correlator synthesizes the sound energy reached by each path, thus highlighting the main correlation peak and reducing the correlation sidelobe. It is easy to use an appropriate threshold to eliminate the multi-path delay ambiguity, so the time resolution is significantly improved compared with the copy correlator. The copy correlation output in Fig. 5.18 has almost two equal peaks. Therefore, in order to avoid communication errors caused by multi-path

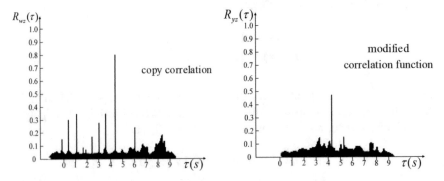

Fig. 5.17 The copy correlation and modified correlation function of deep sea channel (250 N mile)

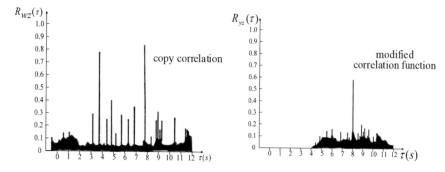

Fig. 5.18 The copy correlation and modified correlation functions of deep sea channel

effect during coded communication, the communication rate is limited by the multi-path delay spread of the channel. The modified correlation technology can increase the communication rate by at least one order of magnitude.

5.5 Response of Adaptive Correlator in Slow Time-Varying Coherent Multipath Channel

The previous chapters of this book, especially the previous sections of this chapter, illustrate that for forward transmission, the marine acoustic channel is a partially coherent channel. In essence, it is a slow time-varying coherent multipath channel. Under this condition, the best time-domain processor is an adaptive correlator. Since the channel is slowly time-varying, the optimal processor must be adaptive. It must perceive the slowly changing information of the channel in real time and adjust the processor to make the best estimation in real time. If the input signal-to-noise ratio is very high, the inter pulse correlator described in Sect. 5.2 is the appropriate optimal processor. When the signal clutter ratio is low, the processing effect of inter pulse correlator is not satisfactory, because serious interference will deteriorate the reference signal quality of correlator, resulting in serious correlation loss. At this time, the modified correlator introduced in Sect. 5.3 will have good effect. This method needs to solve the impulse response function of channel in real time, so as to perceive the time-varying information of channel in real time, so that it can make full use of the coherent component energy realized by different kinds of ways. This method needs to solve the Eq. (5.16) in real time, which should meet the requirement of high computer operation speed. At present, it is difficult and expensive to design such a high-speed computer.

The adaptive correlator introduced in Sect. 3.5 simply realizes the real-time solution Eq. (5.16) by using the L.M.S iterative algorithm. The weight vector W of the adaptive transverse filter is the best estimation of the channel impulse response vector h. Equation (3.34) points out that the frequency response function of the adaptive transverse filter is actually the best estimation of the channel frequency response function, so its Z-transform, that is, the weight vector of the adaptive filter is undoubtedly the best estimation of the channel impulse response vector. Both modified correlator and adaptive correlator make full use of the coherent energy of multi-path arrival signals, so they have high processing gain. The detection problem of active sonar is double selection detection, that is, the processor must judge one of the two: only interference at the input or interference plus target signal at the input.

In fact, the reference signal of adaptive correlator is quite different when there is target signal and no target signal (only interference) at the input. When the signal and interference are independent of each other and there is only interference at the input, the output $y(t)$ of the ideal adaptive matched filter (see Fig. 3.13) will be zero, and $y(t)$ is the reference signal of the adaptive correlator (see Fig. 3.14). Therefore, the output of the adaptive correlator will be zero when there is only interference at the input, which will lead to the infinite output signal-to-noise ratio of the adaptive correlator. In fact, the cross-correlation between target signal and interference is just small instead of being zero, so they are not completely independent of each other, and the output signal-to-noise ratio of adaptive correlator will never become infinite. It shows that another important reason for the high processing gain of the adaptive correlator is that it cannot only adapt to the multi-path coherent channel, but also has a good ability to suppress interference. This section will analyze the processing gain of the adaptive correlator in the slow time-varying coherent multipath channel, taking into account the cross-correlation of transmitted signals and interference, and introduce the offshore experimental results [2].

When the target signal appears at the input and the adaptive matched filter reaches the steady state, that is, when the adaptive learning process ends, the transmission function of the adaptive filter can be seen from Eq. (3.34):

$$H_0(\omega) = \frac{S_{zd}(\omega)}{S_{zz}(\omega)} = H(\omega, t) + \frac{S_{zn}(\omega)}{S_{zz}(\omega)} \qquad (5.26)$$

where $S_{zd}(\omega) = E\{Z^*(\omega) \cdot D(\omega)\}$—cross power spectrum of reference signal and desired signal.

$S_{zz}(\omega) = E\{Z^*(\omega) \cdot Z(\omega)\}$—reference signal power spectrum.

$S_{zn}(\omega) = E\{Z^*(\omega) \cdot N(\omega)\}$—cross power spectrum of reference signal and interference.

$Z(\omega), N(\omega), D(\omega)$—frequency of transmission signal $z(t)$, interference $n(t)$ and desired signal $d(t)$, and:

$$D(\omega) = z(\omega)H(\omega, t) + N(\omega). \qquad (5.27)$$

As can be seen from Eq. (5.26), when the reference signal $z(t)$ (copy of transmitted signal) has a certain correlation with the interference $n(t)$, that is, when the cross power spectrum of the reference signal and the interference is not zero, the system characteristics of the adaptive matched filter will be different from the system characteristics of the channel; When the signal and interference are independent of each other, $S_{zn}(\omega)$ will be zero, and the system characteristics of the adaptive matched filter will be consistent with the system characteristics of the channel. In fact, even if the two are independent of each other, because the integration time length of the adaptive correlator is limited, which is equal to the signal pulse width, the adaptive correlator will make serious errors in the estimation of the cross-correlation function of the reference signal and interference, that is, there will be large errors in the estimation of the cross-correlation power spectrum of the signal and interference. Therefore, in a finite period of time, the first estimation of the cross power spectrum will not be zero, but take a smaller value. It can be expected that when the integration time increases (and therefore the pulse width of the signal must also increase), the estimation error will be reduced, and the primary estimation value of the cross-correlation between the independent signal and the interference will be reduced. It is appropriate to increase the signal pulse width and integration time as much as possible within the allowable range of hardware scale and operation speed. When reverberation becomes the main background interference, it can be expected that the cross-correlation between reference signal and interference will increase slightly (compared with the cross-correlation between signal and noise).

When there is a signal at the input of the adaptive correlator, the output of the correlator is:

$$S_{yd}(\omega)\big|_{s+n} = E\{Y^*(\omega) \cdot D(\omega)\}$$

$$= E\left\{Z^*(\omega)\left(H(\omega) + \frac{S_{zn}(\omega)}{S_{zz}(\omega)}\right)^* \cdot (Z(\omega)H(\omega) + N(\omega))\right\}$$

$$S_{ya}(\omega)\big|_{s+n} = E\left\{Z^*(\omega)Z(\omega)\left(H(\omega) + \frac{S_{zn}(\omega)}{S_{zz}(\omega)}\right)^* \cdot \left(H(\omega) + \frac{Z^*(\omega)N(\omega)}{Z^*(\omega)Z(\omega)}\right)\right\}$$

$$= S_{zz}(\omega)\left|H(\omega) + \frac{S_{zn}(\omega)}{S_{zz}(\omega)}\right|^2 \tag{5.28}$$

When there is only interference at the input of adaptive correlator, the output of correlator is:

$$S_{yd}(\omega)\big|_n = E\{Y^*(\omega)D(\omega)\}$$

$$= E\left\{Z^*(\omega)\frac{S_{zn}^*(\omega)}{S_{zz}(\omega)} \cdot N(\omega)\right\} = |S_{zn}(\omega)|^2 \big/ S_{zz}(\omega) \tag{5.29}$$

According to Eqs. (5.28) and (5.29), the output power signal-to-noise ratio of the adaptive correlator is:

$$(S/N)_{adaptive} = \frac{S_{yd}(\omega)\big|_{s+n}}{S_{yd}(\omega)\big|_n} = \left|\frac{S_{zz}(\omega)}{S_{zn}(\omega)}\right|^2 \cdot \left|H(\omega) + \frac{S_{zn}(\omega)}{S_{zz}(\omega)}\right|^2 \tag{5.30}$$

For signals with limited bandwidth, the total gain can be estimated only by integrating Eq. (5.30) in the frequency domain. For example, for narrowband signals:

$$(S/N)_{adaptive} = \frac{\int S_{zz}(\omega)\left|H(\omega) + \frac{S_{zn}(\omega)}{S_{zz}(\omega)}\right|^2 d\omega}{\int \frac{|S_{zn}(\omega)|^2}{S_{zz}(\omega)} d\omega} \tag{5.31}$$

For narrow-band signals, especially for linear frequency modulation pulses, the shape of its power spectrum function is approximately rectangular, so $S_{zz}(\omega)$ is almost constant in the effective integration interval, and it is noted that the term related to $S_{zz}(\omega)$ can be ignored in the molecule, so there is:

$$(S/N)_{adaptive} \approx \frac{S_{zz}(\omega) \int S_{zz}(\omega)|H(\omega)|^2 d\omega}{S_{zz}(\omega) \int \frac{|S_{zn}(\omega)|^2}{S_{zz}(\omega)} d\omega} = \frac{\int |S_{zz}(\omega) \cdot H(\omega)|^2 d\omega}{\int |S_{zn}(\omega)|^2 d\omega} \tag{5.32}$$

Apply the power theorem to Eq. (5.32), that is, the energy of the signal is equal to the frequency domain integral of its power spectrum and the peak value of the correlation function. Therefore, Eq. (5.32) can be rewritten as:

$$(S/N)_{adaptive} \approx \left(\frac{R_{zz}(\tau) * R_n(\tau, t)|_{\tau=0}}{R_{zn}(\tau)|_{\tau=0}} \right) \tag{5.33}$$

When the output signal-to-noise ratio is expressed in dB, Eq. (5.33) shall be rewritten as:

$$(S/N)_{adaptive} \approx 20 \lg \left(\frac{R_{zz}(\tau) * R_n(\tau, t)|_{\tau=0}}{R_{zn}(\tau)|_{\tau=0}} \right) \tag{5.34}$$

As can be seen from Eqs. (5.30), (5.31) and (5.34), when the signal and interference are not correlated with each other, the processing gain of the adaptive correlator tends to infinity. In fact, due to the above reasons, and because the word length and the number of weights of the adaptive canceller are limited, $R_{zn}(\tau)$ will not be zero. The processing effect can be brought into full play only when the parameters are reasonably selected and the adaptive correlator is carefully designed; Also, when the input signal-to-noise ratio decreases, the processing gain will decrease accordingly.

For a practical adaptive correlator, weight noise, quantization noise and learning speed have an important impact on its performance. The learning speed of the adaptive canceller should be much faster than the time-varying rate of the channel, that is, the time coherence length of the channel should be much greater than the adaptive learning time length. Generally speaking, the speed of adaptive learning is very fast, so the above conditions can be met.

The lake test results of the adaptive correlator are described in Sect. 3.6, and now the sea test results are introduced. Active sonar is installed on surface ships. Its detection target is a medium-sized submarine sailing at low speed. The surface ship approaches the target at a speed of 18kn, and the sonar transmitted signal is linear frequency modulation pulse with a pulse width of 0.5 s and a frequency modulation width of 500 Hz. The transmitted electrical signal and the received echo signal are recorded on two-channel magnetic tape respectively. During the experiment, the target distance is about 5 km, and the water depth in the test sea area is about 40 m. The laboratory test shows that the main interference background is reverberation, and the target echo is very strong, that is, it has a high signal-to-mixing ratio.

We conduct copy correlation and adaptive correlation analysis on the audio tapes of sonar sea test in the laboratory. The analysis system is shown in Fig. 5.19. The purpose of using mixer is to reduce the required sampling frequency and computation. The data is sampled at a sampling rate of 2 kHz, and the JEF-16AN analysis system is used for copy correlation and adaptive correlation analysis. The results are shown in Tables 5.1 and 5.2.

The data listed in the table are the average results of 26 samples, and all signal-to-noise ratios are defined by power signal-to-noise ratio. Although the target echo

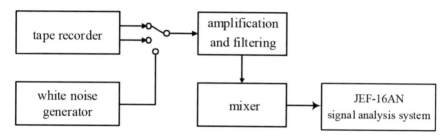

Fig. 5.19 The diagram of signal analysis system

Table 5.1 Comparison of output SNR of adaptive correlator and copy correlator

Input signal-to-noise ratio (dB)		Interference of reverberation				Interference of noise			
		10.3	5.0	0	−5.0	10.0	5.0	0	−5.0
Output signal-to-clutter ratio (dB)	Copy	17.5	12.3	8.0	3.5	15.8	10.5	5.9	1.9
	Adaptive	25.8	19.7	13.5	6.4	25.0	18.8	11.4	5.3
	Subtraction	8.3	7.4	5.5	2.9	9.2	8.3	5.5	3.4

Table 5.2 Comparison of normalized correlation coefficients of target echo between adaptive correlator and copy correlator output

Input signal-to-mixing ratio (dB)	The normalized adaptive correlation coefficient of the target echo from copy correlator output	The normalized adaptive correlation coefficient of the target echo from adaptive correlator	The ratio of the two (dB)
17.5	0.32	0.53	4.4

signal recorded on the tape also contains reverberation components, the reverberation intensity is much smaller than the echo intensity, so it can be regarded as a "clean" target echo signal. The samples with different signal-to-mixing ratio are artificially made in the computer, that is, the adjacent reverberation in front of the target echo is amplified and the clean target echo is added in the computer to make the samples with low signal-to-mixing ratio. The noise samples are sent to the computer by the white noise generator through the system shown in Fig. 5.19, and samples with different signal-to-noise ratios are made in the same way as above.

The results in Table 5.1 show that the processing gain of the adaptive correlator is about 3–9 dB higher than that of the copy correlator under 18 kn Doppler conditions. If it is assumed that the output of the correlator is detectable when it has a signal-to-noise ratio of 6 dB, the detection threshold of the adaptive correlator is about 5 dB lower than that of the copy correlator. Experiments show that when the input signal-to-noise ratio decreases, the processing gain of the adaptive correlator also decreases, which is consistent with the theoretical prediction.

Table 5.2 shows that the normalized adaptive correlation coefficient of the target echo is about 4.4 dB higher than the normalized copy correlation coefficient, which shows that the adaptive correlator has good adaptability to the slow time-varying multi-path coherent channel.

5.6 System Function of Moving Sound Source in Coherent Channel

As mentioned earlier, when only considering the forward propagation of sound wave, and when the sound source and receiver are fixed, the energy of incoherent scattering is actually unimportant, and the sound channel can be regarded as a slow time-varying coherent multipath channel. Generally speaking, the sonar carrier and target are moving without losing generality. It can be assumed that the receiver is stationary and the sound source is moving. Can the moving sound source channel be regarded as a slowly time-varying coherent multipath channel? Are the various coherent processors still valid when the target is moving? The following part of this chapter will answer the above questions.

When there is relative motion between the sound source and the receiver, there will be no stable coherent component in the channel, and the system function of the channel will be time-varying. We still assume that the channel is coherent and multipath. We discuss the time-varying system function of the channel from the perspective of ray acoustics and study the time-varying rate of the channel when the sound source moves. We do not expect to establish a practical model for predicting the channel system function of the moving sound source, but try to understand the relationship between the time-varying of the channel and the motion of the sound source under the simplified model, and the magnitude of channel time-varying rate is predicted.

When reviewing the coherent function of the multi-channel, it can be seen from Eq. (3.6) that the multi-channel response function of the system is the static impulse function $h(t)$:

$$h(t) = \sum_{i=1}^{N} A_i \delta(t - \tau_{0i}) \tag{5.35}$$

The impulse response function of the channel should be understood as the receiving waveform of the receiving point when the point sound source radiates δ pulse. In Eq. (5.35), the received signal is regarded as the superposition reached through N rays. The signals transmission along each sound path are colorless and scattered. A_i and τ_{0i} are determined by the environmental conditions of the channel, and their values can be obtained through ray calculation. Making Fourier transform on Eq. (5.35), we can obtain the transmission function of the channel:

$$H(\omega) = \sum_{i=1}^{N} A_i e^{-j2\pi f \tau_{0i}} \tag{5.36}$$

When the sound source is moving, the impulse response function of the channel can be obtained from the modified Eq. (5.35), which is:

$$h(t, t_0) = \sum_{i=1}^{N} A_i(t_0) \delta\left[t - \gamma_i\right] \tag{5.37}$$

t_0 is the starting time of observation, t is the actual observation time, and γ_i is the transmission time of the acoustic signal from the sound source to the receiving point along the ith path. Generally speaking, when the sound source moves, the signal amplitude $A_i(t_0)$ reached by each path also changes, but compared with γ_i, $A_i(t_0)$ is a slow change function of time t. Here, the multi-path interference of the sound field is more sensitive to the multi-path phase relationship. Therefore, it is assumed that the amplitude $A_i(t_0)$ is independent of time t within a long observation time interval. The system function of the channel can be determined only by determining the time function form of the propagation delay γ_i.

If the pulse of route i is received at time t, the time when the δ pulse is emitted at the static sound source shall be $(t - \tau_{0i})$, τ_{0i} is the propagation delay through route i, and now the sound source is moving. Considering the Doppler phenomenon, the δ pulse passing through the same route i shall be $K_i(t - \tau_{0i})$ at the emission time of the moving sound source, and K_i should be called the Doppler coefficient. Transmission delay should be the difference between transmitting time and receiving time:

$$\gamma_i = t - K_i(t - \tau_{0i}) \tag{5.38}$$

Substituting Eq. (5.38) into Eq. (5.37), the impulse response function is:

$$h(t, t_0) = \sum_{i=1}^{N} A_i(t_0)\delta\{t - [t - K_i(t - \tau_{0i})]\} \tag{5.39}$$

Next, the Doppler coefficient K_i of the moving sound source is determined. Omitting the foot mark I, the propagation delay r along a certain path shall be determined by Eq. (5.38):

$$\tau = t - K(t - \tau_0) \tag{5.40}$$

When the sound source moves, the propagation delay τ changes with time t, and its change rate can be obtained by deriving from Eq. (5.40):

$$\frac{d\tau}{dt} = 1 - K - (t - \tau_0)\frac{dK}{dt} \tag{5.41}$$

Also:

$$\frac{d\tau}{dt} = -\frac{\upsilon L_\theta}{c(r, t)} \tag{5.42}$$

where υ is the motion velocity vector of the sound source, L_θ is the unit tangent vector of the ray from the sound source emitting point, and $c(r, t)$ is the sound velocity at the sound source. The numerator of Eq. (5.42) represents the projection component of the sound source velocity in the tangent direction of the ray, and the time delay change rate is obtained by dividing it by the sound velocity. The sweep angle of the ray is θ. Expanding the dot product in Eq. (5.42), we have:

$$\frac{d\tau}{dt} = -\frac{\frac{dR}{dt} \cdot \cos\theta + \frac{dz}{dt}\sin\theta}{c(r, t)}$$

$\frac{dR}{dt}$ is the horizontal component of the motion speed of the sound source in the direction of its connection with the receiving point, $\frac{dz}{dt}$ is the vertical component of the speed, that is, the heave speed of the sound source, θ and $\frac{dz}{dt}$ are both small, so the above equation can be approximate to:

$$\frac{d\tau}{dt} \approx -\frac{dR}{dt} \cdot \frac{\cos\theta}{c(r, t)} \tag{5.43}$$

According to Snell's law, $\frac{\cos\theta}{c(r,t)}$ is constant along the ray. The motion of the sound source is often random. Even if the working condition of the ship remains unchanged and the ship tries to keep straight-line navigation, due to the influence of waves and steering with a fixed course, it can always be regarded as the sum of the motion speed of the sound source as a constant and a random small quantity, so that:

$$\frac{dR}{dt} = v_0 + \sigma_v \xi \tag{5.44}$$

σ_v is the standard deviation of the random velocity component, and ξ is the normalized random variable with zero mean, that is, the variance of ξ is 1. Substituting Eq. (5.44) into Eq. (5.43), we have:

$$\frac{d\tau}{dt} \approx -\frac{1}{c}[v_0 \cos\theta + \sigma_v \cdot \cos\theta \cdot \xi] = -K_0^{-1} - \varepsilon\xi + 1 \tag{5.45}$$

where $\varepsilon = \frac{\sigma_v \cdot \cos\theta}{c}$

$$K_0^{-1} = \left(1 - \frac{v_0}{c} \cdot \cos\theta\right)^{-1}, \quad \frac{v_0}{c} << 1 \tag{5.46}$$

Since $\frac{\cos\theta}{c(r,t)}$ is a constant, it has no relation with K_0 and t. Substitute Eq. (5.45) into Eq. (5.41), and there is:

$$\frac{d[(t - \tau_0)K]}{dt} = K_0^{-1} + \varepsilon\xi$$

The integral of the above differential equation is:

$$K = \frac{1}{t - \tau_0}\left\{K_0^{-1}(t - \tau_0) + \varepsilon \int_0^{t-\tau_0} \xi(x)dx\right\}$$

$$= K_0^{-1} + \frac{\varepsilon}{t - \tau_0} \int_0^{t-\tau_0} \xi(x)dx \tag{5.47}$$

According to Eqs. (5.39) and (5.47), the impulse response function of the moving sound source channel is

$$h(t, t_0) = \sum_{i=1}^{N} A_i(t_0)\delta[t - (t - K_i(t - \tau_{0i}))] \tag{5.48}$$

$$K_i = K_{0i}^{-1}\left[1 + \frac{K_{0i}\varepsilon_i}{t - \tau_{0i}} \int_0^{t-\tau_{0i}} \xi(x)dx\right] \tag{5.49}$$

Fourier transform Eq. (5.48) to obtain the transmission function of the channel as follows:

$$H(f, t) = \sum_{i=1}^{N} A_i(t_0) e^{-j2\pi f \left[\tau_{0i} - (K_i - 1)(t - \tau_{0i}) \right]} \qquad (5.50)$$

Equation (5.50) shows that the transmission function of the moving sound source channel is related to t_0, especially the phase factor $2\pi f \tau_{0i} = 2\pi f \times \tau_{0i}(t_0)$ is sensitive to t_0, which is equivalent to that when the position of the sound source changes significantly due to the movement, the position change will significantly change the phase relationship of the multi-path arrival ray cluster, so as to change the interference of sound waves in each path, resulting in the change of the shape of the transmission function. The change rate of the transmission function caused by this depends on the moving speed of the sound source and the scale of the spatial pattern of the sound field interference. The scale of the spatial pattern of the sound field interference in the deep sea is large, so the change rate of the transmission function of the moving sound source channel is slow, and the time-varying rate in the shallow sea channel is much faster. Equation (5.50) synthesizes the influence of Doppler on channel transfer function, and makes numerical calculation in order to estimate the time-varying rate of channel caused by Doppler. Taking the deep-sea ray parameters given in Fig. 3.2 and Table 3.1, for the sound source motion speed of 10 kn and frequency of 33 Hz, calculate the time-varying transmission function of the channel according to Eq. (5.50). Assuming the ray parameters do not change within the observation time (only to investigate the influence of Doppler), the amplitude time-varying characteristics and phase time-varying characteristics of the transmission function are shown in Fig. 5.20, which shows that the time-varying rate of the channel is about 5 min, and the time-varying effect of the channel caused by Doppler is slow.

Fig. 5.20 Time varying effect of Doppler induced transfer function (sound source speed: 10 kn, $f = 33$ Hz; ray parameters: see Table 3.1)

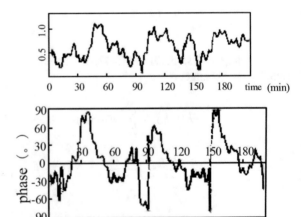

5.7 Loss of Cross-Correlation During Target Motion

When the sonar carrier and the target move relative to each other, the signal waveform at the receiving point will be compressed or stretched in time compared with the radiation waveform. The former corresponds to the condition when they are close to each other and the latter corresponds to the condition when they are far away from each other. This phenomenon is called Doppler phenomenon, which will lead to the decrease of correlation number, called cross-correlation loss.

As shown in Fig. 5.21, it does not lose generality. Set the receiving hydrophone at point R, the point sound source S moves at speed v, and ψ is the velocity vector and the angle between the sound source and the connecting line between the receiving point. If the radiation waveform of the sound source is $z(t)$, the received signal $\omega(t)$ based on the Doppler principle is:

$$\omega(t) = z[K(t - \tau_0)] \tag{5.51}$$

where τ_0 is the transmission delay from the sound source to the receiving point R when the sound source is stationary with K as the Doppler coefficient. If the amount of motion is set to zero in Eq. (49.5):

$$\left.\begin{array}{l} K = K_0^{-1} = \left(1 - \frac{v}{c} \cdot \cos \psi\right)^{-1} \\ k \approx 1 + \frac{v}{c} \cdot \cos \psi = 1 + \delta \\ \delta = \frac{v}{c} \cdot \cos \psi \end{array}\right\} \tag{5.52}$$

where c represents the sound velocity, δ denotes the Doppler coefficient.

If the frequency of the single frequency harmonic sound wave radiated by the sound source is f, the frequency of the received sound wave shall be $f + \Delta f$, and it can be seen from Eq. (5.50):

Fig. 5.21 Relative geometric relationship between moving sound source and receiving point

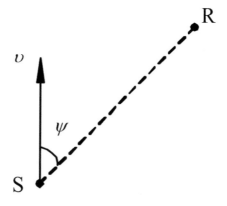

Fig. 5.22 Relative
geometric relationship
between target and array

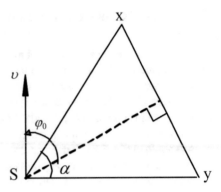

$$\Delta f = \frac{\upsilon}{c} \cdot \cos \psi \cdot f \tag{5.53}$$

That is, if the sonar and the target move close to each other, the received sound wave frequency will be higher than the radiated sound wave frequency, otherwise, the received frequency will become lower.

Figure 5.22 shows the relative geometric relationship between the hydrophone array (x, y) and the target. If the viewing angle of the array to the target is a, the Doppler coefficient δ is:

$$
\begin{aligned}
\delta_x &= \frac{\upsilon}{c} \cdot \cos\left(\varphi_0 - \frac{a}{2}\right) \\
\delta_y &= \frac{\upsilon}{c} \cdot \cos\left(\varphi_0 + \frac{a}{2}\right)
\end{aligned}
\tag{5.54}
$$

Because the target moves relative to the hydrophone array, the Doppler effect makes the signal waveforms received by hydrophones x and y different, so the correlation number of the signals received by the two hydrophones will be reduced. According to Eq. (5.51):

$$
\begin{aligned}
\omega_x(t) &= z[(1 + \delta_x)t] \\
\omega_y(t) &= z\big[(1 + \delta_y)t\big]
\end{aligned}
\tag{5.55}
$$

Only the position sensitive to Doppler effect is investigated, that is, as shown in Fig. 5.22, when the sound source is located on the vertical line of the array line, δ_x and δ_y are given by Eq. (5.54). The unimportant fixed delay is omitted in Eq. (5.55). The cross-correlation function is:

$$
\begin{aligned}
R_{xy}(\tau) &= \langle z[(1 + \delta_x)t]z^*\big[(1 + \delta_y)(t + \tau)\big]\rangle \\
&= \langle z(t)z^*\big[(1 + \delta_{xy})(t + \tau)\big]\rangle
\end{aligned}
\tag{5.56}
$$

where

$$\delta_{xy} = \delta_y - \delta_x = -2\frac{v}{c}\sin\varphi_0 \sin\frac{a}{2} \tag{5.57}$$

Equation (5.56) is applicable to any radiation signal waveform in infinite space. In order to give the quantitative concept of cross-correlation loss caused by target motion, it is advisable to investigate the harmonic sound wave radiated by the sound source, and replace the ensemble average with time average. It can be seen from Eq. (5.56):

$$R_{xy}^T(0) \triangleq \frac{1}{T}\int_0^T \cos\omega t \cdot \cos\omega(1+\delta_{xy})t\,\mathrm{d}t$$

$$\approx \frac{2}{T}\int_0^T \cos(2\pi f\delta_{xy}t)\,\mathrm{d}t = \frac{2\sin 2\pi f\delta_{xy}T}{2\pi f\delta_{xy}T} \tag{5.58}$$

where T is the integration time length. The equation shows that for large-scale arrays, the cross-correlation loss caused by Doppler cannot be ignored, and the correlation integration time must be limited to avoid intolerable correlation loss. Through the following numerical examples, we can give the quantity concept that is more in line with the actual situation. If the array length of passive sonar is 40 m and the target is 2 km in front of the array, then $a = 0.02$ rad, $\varphi_0 = 90°$. If the target angular velocity (or the angular velocity of the carrier of the array) is $\Omega = 2°/s$ and the acoustic velocity is $c = 1450$ m/s, then:

$$\delta_{xy} = -0.0009629$$

$$R_{xy}(0, T) = \frac{\sin 2\pi f \times 0.0009629T}{2\pi f \times 0.0009629T}$$

When $f = 2.8\text{kHz}$, the allowable correlation loss is 3 dB, the maximum allowable integration time length can be obtained from the above equation is $T = 0.083$ s.

When the background interference received by the two hydrophones is independent of each other, the longer the integration time, the higher the processing gain of the correlator, but the time length of coherent integration is limited when the target moves. Therefore, in order to further improve the processing gain of the system, post-processing must be adopted, that is, an envelope detector must be connected behind the output of the correlator, then smooth the output envelope again with RC filter or first-order recursive filter. Since the target is moving, the peak position of the correlation function output by the correlator will also move on the time delay axis, so the length of the post integration time is also limited. Figure 5.23 illustrates the above reason. The envelopes 1, 2 and 3 of the correlation functions represent the output of the adjacent discrete-time correlator respectively. It is not difficult to understand from the figure that the post integration time length should not be too

Fig. 5.23 The movement
diagram of correlation peak
during target movement

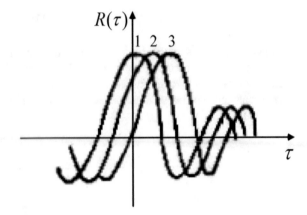

large. A reasonable selection should ensure that the main lobes of the correlation
peaks should not be separated from each other within the integration time length.

The spatiotemporal correlation function [3] in the plane acoustic field of limited-
band white noise is:

$$R_\omega(d, \tau) = \frac{\sin \frac{\omega_2 - \omega_1}{2}(\tau + \tau_2 - \tau_1)}{\frac{\omega_2 - \omega_1}{2}(\tau + \tau_2 - \tau_1)} \cos \left[\frac{\omega_2 + \omega_1}{2}(\tau + \tau_2 - \tau_1) \right] \tag{5.59}$$

where $(\tau_2 - \tau_1)$ is the delay difference of the signal received by the two elements
of the array, ω_2, ω_1 are the angular frequency of the upper and lower limits of the
signal, and d is the distance between the two hydrophones. And:

$$\tau_2 - \tau_1 = \frac{d \sin \theta}{c} \tag{5.60}$$

where θ is the angle between the incident sound wave and the normal of the array.
The first factor in Eq. (5.59) describes the envelope of the correlation function. If
the correlation function envelopes of the two extremes within the integration time
length are overlapped at 0.7, the first factor of Eq. (5.59) shows that the maximum
integration time length can be determined by the following equation:

$$T \approx \frac{2c}{\Omega d(f_2 - f_1)} \tag{5.61}$$

where Ω is the angular velocity of target motion in rad/s; f_2, f_1 is the upper and lower
frequency respectively. For the above example of passive sonar, if $f_2 - f_1 = 2$ kHz,
the allowable integration time constant is 1 s.

References

1. Fink M, et al. Acoustic time-reversal mirrors. Inverse Probl. 2001;17:1–38.
2. Hui JY, Wang LS. Adaptive matching filter and adaptive correlator. Underwater Acoustic Communication; 1986.
3. Urick RJ. Principles of underwater sound, 3rd ed., vol. 22. Peninsula Publishing Los Altos, California; 1983. p. 23–24.

Chapter 6
Reverberation Channel

There are many kinds of inhomogeneities in seawater medium and boundary. Sound waves encounter these inhomogeneities and produce scattering. All scattered sound waves reaching the receiving point at the same time stack to form reverberation. The reverberation is received immediately after the sonar signal is transmitted. It sounds like a long, slowly weakening, vibrating, and fluctuating sound. Reverberation is a unique interference of active sonar. The operating range of long-range high-power sonar is limited by reverberation, and the operating range of special sonars such as mine detector and fish detector are also limited by reverberation. Suppressing reverberation and detecting signal in reverberation background is one of the important topics in active sonar detection technology. Reverberation is sometimes used to study media, such as classifying marine biological resources, predicting the size of fishing season and studying the characteristics of seabed. Side sonar uses reverberation to map the seabed topography, and Doppler sonar uses reverberation to measure the sailing speed of ships. Reverberation is not only an important interference, but also a useful signal, depending on different applications.

6.1 Average Characteristics of Reverberation

Firstly, the average energy characteristics of reverberation are briefly analyzed. In the design and application of active sonar, it is very necessary to estimate the reverberation level of sonar system under predetermined working conditions. The reverberation level can be expressed as:

$$RL = SL - 2TL_R + S_{s,v} + 10 \log A, V \tag{6.1}$$

The physical meaning of Eq. (6.1) will be explained at the end of this section. First, we will explain the physical meaning of each quantity in the equation.

© Harbin Engineering University Press 2022
J. Hui and X. Sheng, *Underwater Acoustic Channel*,
https://doi.org/10.1007/978-981-19-0774-6_6

   In the equation, SL is the sound source level. The reference sound pressure is taken as $1\mu P_a$. $2TL_R$ is transmission loss of reverberation; $S_{s,v}$, A, and V are the area and volume scattering intensity, and the scattering area or scattering volume. According to $S_{s,v}$, which is measured in advance in the experiment, selecting the appropriate value and using the ray theory, the reverberation level prediction can be completed quickly on the digital computer.

   Reverberation is regarded as the superposition of all scattered waves arriving at the receiving point at the same time. The intensity of the scattered wave is directly proportional to the intensity of the projected wave. Therefore, the reverberation level is directly proportional to the sound source level. The reverberation intensity increases with the increase of transmission power. Increasing transmission power does not improve the signal-to-mixing ratio; Scattering area A or scattering volume V is an increasing function of beam width and signal width. Reducing beam width and signal width will reduce the reverberation intensity. In addition, the reverberation level also has some relationship with the signal waveform.

   Reverberation can be divided into sea surface, seabed and volume reverberation according to the spatial distribution type of scatterers. The scattering intensity $S_{s,v}$ is equal to the ratio of the scattering sound intensity at 1 m per unit scattering area or volume to the incident plane wave sound intensity, taking its decibel value. Figure 6.1 shows the relationship between volume scattering intensity and depth measured experimentally in deep sea. The layer with large scattering intensity is called deep-water scattering layer. Except for the deep-water scattering layer, the volume scattering intensity decreases with the increase of depth. The deep-water scattering layer is composed of biological scatterers. Its depth varies day and night. It lives deep during the day and shallow at night. The depth changes most violently at sunrise and sunset, and the depth changes as much as hundreds of feet. Therefore,

Fig. 6.1 An example of the relationship between volume scattering intensity and depth

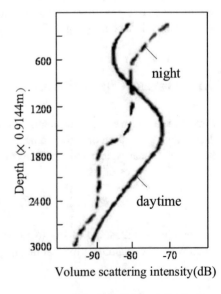

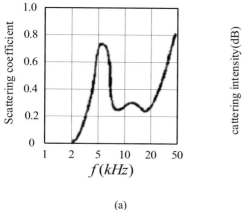
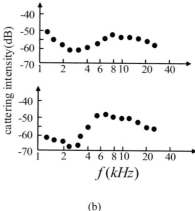

(a)   (b)

Fig. 6.2 The frequency characteristic example of scattering intensity of deep water scattering layer. **a** Measurement results of Areeba of the Soviet Union in the South China Sea. **b** McElroy's measurements in the North Atlantic

the volume scattering intensity near the sea surface changes by 10 dB at sunset and sunrise, which is enough to make the sonar performance of long-range surface ships very poor at night. The deep-water scattering layer not only seriously interferes with the sonar working in the convergence area, but also has important interference with the bathymeter working in the deep sea, especially the digital bathymeter using the amplitude threshold detector. If it is not designed properly, the bathymeter will mistake the deep-water scattering layer as "seabed".

The scattering intensity of the deep-water scattering layer is −70 to −55 dB in the frequency band of 1–2 kHz, with obvious frequency characteristics, as shown in Fig. 6.2. Marine organisms can usually be classified according to their frequency response characteristics.

Figure 6.2a is an example measured by Soviet scholars in the South China Sea. The peak value of scattering intensity is near 5 kHz; Figure 6.2b shows the measurement results of American scholars. Each curve is the average value of multiple measurements at several stations (with roughly the same frequency characteristics). Although the frequency characteristics measured by each station are different from each other to varying degrees, they all have the following common characteristics:

1. From 1 to 3.15 kHz, the volume scattering coefficient decreases with the increase of frequency, the minimum value is scattered between 2 and 4 kHz, and the most likely value is 3.15 kHz.
2. There is a peak between 4 and 8 kHz, and the most likely frequency is 6.3 kHz. The difference between the peak value and the minimum value is often more than 10 dB. The second peak is about 12.5 kHz.

Therefore, when the active sonar frequency for remote operation is selected near 3 kHz, the volume reverberation interference will be small, and it is inappropriate to select the operating frequency around 5 kHz.

The sea surface scattering intensity is a function of wind speed, grazing angle, and frequency. The experimental results are shown in Fig. 6.3. It is generally considered that a layer of bubble adjacent to the sea surface is the main scattering source in the case of small grazing angle; At medium grazing angle, the main cause of scattering is the uneven sea surface; For the case of large grazing angle, the small mirror of the sea surface is the main scattering source.

The seafloor scattering intensity is related to the grazing angle, frequency, and seafloor sediment. Their relationship is shown in Fig. 6.4. The seafloor scattering intensity increases rapidly with the increase of grazing angle. Therefore, for short-range sonar with narrow beam, such as mine detector, fishing detector, Doppler sonar, side sonar, collision avoidance sonar, etc., the directional side lobe of the transducer array should be designed as low as possible, so as to reduce the strong seabed reverberation interference, which has a very important impact on the performance of the equipment. Because the scattering intensity of the seabed decreases faster than the

Fig. 6.3 The relationship between sea surface scattering intensity and sweep angle and frequency. − Chapman: wind velocity is 20 kn. O Urik: wind velocity is 17.5 kn. + Garison: wind velocity is 14–30 kn

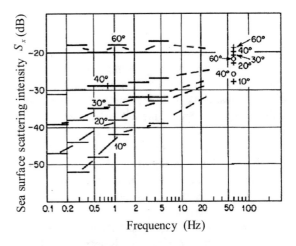

Fig. 6.4 The relationship between seabed scattering intensity and grazing angle

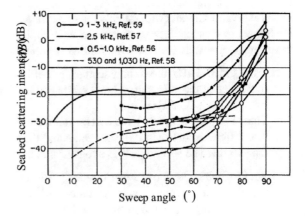

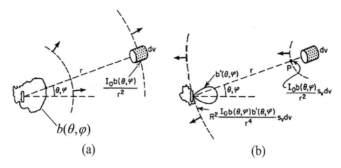

Fig. 6.5 Volume scattering geometry. **a** Launch. **b** Receive

first power law at small grazing angle, the long-distance shallow sea signal mixing ratio will be a non-decreasing function of distance.

Reverberation is a non-stationary random process, and its mean and variance change with time. The typical dynamic range is 50–70 dB within 10 s. This is one of the important reasons why the effect of many signal processing technologies is not ideal, resulting in a large false alarm rate or uneven background contrast on flat B-type display. Therefore, before the signal enters the processor, the reverberation must be "quasi stabilized", that is, the so-called "dynamic range compression and normalization" technology is used to stabilize the average value or variance of the reverberation envelope of the input processor. "Normalization" technology is also known as "constant false alarm rate" technology. The classical normalization technologies include automatic gain control, time gain control, logarithmic amplifier, etc. Due to the application of computer in sonar, adaptive normalization technology will have a bright future.

Next, let's take volume reverberation as an example to illustrate the physical meaning of Eq. (6.1) and the attenuation law of the average intensity of volume reverberation with time.

As shown in Fig. 6.5, there is a directional transmitting transducer placed in infinite space, and it is assumed that:

1. The ray is a straight and the medium has no absorption, so the attenuation other than spherical wave attenuation is ignored.
2. The distribution of scatterers is uniform everywhere.
3. Only one scattering is considered, and the reverberation caused by multiple scattering can be ignored.

Now let's find the average attenuation law of volume reverberation under the above conditions.

Let $b(\theta, \phi)$ and $b'(\theta, \phi)$ be the transmitting and receiving directivity function (power directivity function) respectively, and the sound intensity 1 m above the sound axis from the acoustic center of the sound source is I_0, that is, the sound source level is $SL = 10 \log I_0$. According to the definition of directivity function, the sound intensity 1 m away from the sound source in the (θ, φ) direction is $I_0 b(\theta, \varphi)$.

A small volume of scatterer dV away from the sound source r in this direction is investigated, and the incident sound wave intensity is $I_0 b(\theta, \varphi)/r^2$. In the direction towards the sound source, the backscattered sound intensity at 1 m away from dV is $[I_0 b(\theta, \varphi)/r^2] \times S_V dV$, where S_V is the volume scattering intensity $S_v = 10 \log S_V$, which is expressed in dB. If the reverberation intensity returned from dV to the sound source is $(I_0/r^4)b(\theta, \varphi)S_V dV$, the receiving directivity function is $b'(\theta, \varphi)$, and the sensitivity of the water intake hearing device is 1, the reverberation intensity generated by receiving DV in this direction is $(I_0/r^4)b(\theta, \varphi)b'(\theta, \varphi)S_V dV$. If dV is subdivided, the contribution of all volume elements dV can be expressed as $(I_0/r^4)S_V \cdot \int_V b(\theta, \varphi)b'(\theta, \varphi)dV$ by integral, so the reverberation level is:

$$RL = 10 \log \left(\frac{I_0}{r^4} S_V \int_V b(\theta, \varphi)b'(\theta, \varphi)dV \right)$$

The volume element dV must be carefully examined. As shown in Fig. 6.6, let dV be a very small cylinder with a finite length, and its bottom surface is perpendicular to the incident direction of sound wave. Therefore, the bottom area of the volume element is $r^2 d\Omega$, in which $d\Omega$ is the solid angle of the volume element dV to the sound source. For a pulsed sound source, the length of dV can be determined by the following method, that is, all scattering caused in dV may return to the sound source (or receiver) at the same moment. This requirement indicates that if an acoustic pulse irradiates the scatterer in dV, it can be assumed that the time when the head of the acoustic pulse is scattered back to the sound source by the scatterer "behind" in dV is the same as the time when the tail of the pulse is scattered back to the sound source (or receiver) by the scatterer "in front" of dV. Therefore, the length of this scattering voxel dV should be $c\tau/2$, in which τ is the pulse width and c is the acoustic velocity. Therefore (Fig. 6.7):

$$dV = r^2 \frac{c\tau}{2} d\Omega$$

Fig. 6.6 The schematic diagram of volume element dV

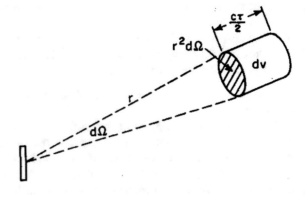

Fig. 6.7 Actual and equivalent directivity pattern

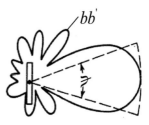

Therefore, the reverberation level is:

$$RL \cong 10 \log \left(\frac{I_0}{r^4} \cdot r^2 \cdot \frac{c\tau}{2} \cdot S_V \cdot \int bb' d\Omega \right)$$

Here we assume $c\tau/2 << r$.

For volume reverberation, the integral $\int_V bb' d\Omega$ can be interpreted as the equivalent beam width of the receiver transmitter directivity combination. The directivity function set in the solid angle ψ is always 1 and zero outside ψ. If such an ideal directivity function is used to replace the actual comprehensive directivity function bb', then ψ is:

$$\int_0^{4\pi} bb' d\Omega = \int_0^{\psi} 1 \times d\Omega = \psi$$

Therefore, the solid angle ψ can be regarded as the solid angle stretched by the ideal directivity pattern. There is a flat response in ψ and zero outside ψ for the response of reverberation, this ideal directivity function is equivalent to the actual comprehensive directivity function.

Then the reverberation level can be expressed as:

$$RL = 10 \log \left(\frac{I_0}{r^4} S_V \frac{c\tau}{2} \psi r^2 \right)$$

The above equation can be expanded as:

$$RL = SL - 40 \log r + S_V + 10 \log V \tag{6.2}$$

$$V = \frac{c\tau}{2} \psi r^2$$

V is called the reverberation volume, that is, the volume occupied by the scatterer producing reverberation at a certain time.

If the transmitting transducer and receiving hydrophone are non-directional, $b(\theta, \varphi) = b'(\theta, \varphi) = 1$, there is:

$$\psi = \int_0^{4\pi} bb' d\Omega = 4\pi$$

According to Eq. (6.2), the average intensity of volume reverberation decreases with distance according to the quadratic law, that is, the quadratic law of time.

6.2   Scattering Function of Reverberation

The reverberation channel is random and unstable in time domain. The scattering intensity has frequency characteristics, which means that the reverberation is not stable in the frequency domain. However, if the observed frequency band is not too wide, it can be considered that the frequency response is flat in the investigated frequency band, so it can be considered as stable in the frequency domain. Although reverberation is unstable in time domain, it changes slowly, so it can be quasi stabilized. The system function of reverberation channel can be written into two parts, namely:

$$\left. \begin{array}{l} R_{hr}\left(\tau', \tau' + \tau; t, t + \Delta t\right) = R'_{hr}\left(\tau', t\right) R''_{hr}(\tau, \Delta t) \\ R_{Hr}(f, f + \Delta f; t, t + \Delta t) = R'_{Hr}(f, t) R''_{Hr}(\Delta f, \Delta t) \end{array} \right\} \qquad (6.3)$$

where the foot mark r represents the statistics of the reverberation process. $R'_{hr}(\tau', t)$ and $R'_{Hr}(f, t)$ represents the slowly changing part of the process, while $R''_{hr}(\tau, \Delta t)$ and $R''_{H}(\Delta f, \Delta t)$ describes the reverberation process that meets the WSSUS conditions. It is not difficult to know from Eq. (6.3) that the reverberation satisfying the generalized stationary condition is:

$$\left. \begin{array}{l} R''_{Hr}(\Delta f, \Delta t) = \left\langle \frac{H_r(f,t) H_r^*(f + \Delta f, t + \Delta t)}{R_{Hr}(f,t)} \right\rangle \\ R''_{hr}(\tau, \Delta t) = \left\langle \frac{h_r(\tau', t) h_r(\tau', + \tau, t + \Delta t)}{R_{hr}(\tau', t)} \right\rangle \end{array} \right\} \qquad (6.4)$$

$\frac{H_r(f,t)}{[R'_{Hr}(f,t)]^{1/2}}$ is the reverberation process after normalization, and "''" indicates the reverberation process after stabilization. Only the reverberation process that has been stabilized is discussed below.

1.   Poisson reverberation process

All scattered signals arrive at the same time. It is assumed that the secondary scattering effect can be ignored, and the waveform of the scattering signal is the same as the transmission waveform. If the scatterer is moving, for the narrow-band signal, the scattering waveform is unchanged except for adding a Doppler shift. Under the above assumptions, the reverberation process can be written as [1]:

$$\omega_r(t) = \sum_{i=1}^{N(t)} A_i(t)e^{j2\pi\varphi_i t}z(t - t_i) \tag{6.5}$$

where $N(t)$ is the total number of scatterers forming reverberation at time t, that is, the scattered waves of $N(t)$ scatterers arrive at the receiving point at the same time at time t, and φ_i represents the Doppler frequency shift caused by the motion of the i-th scatterer. $A_i(t)$ represents the amplitude of the scattering signal of the i-th scatterer received by the receiving point at time t, including two factors: scattering intensity and propagation attenuation.

Assuming that the scatterers are independent of each other, $N(t)$ satisfies Poisson distribution $\rho(N, \tau, \varphi)$ [2]:

$$\rho(N, \tau, \varphi) = \frac{1}{N!}[\rho(\tau, \varphi)\Delta\tau\Delta\varphi]^N \exp[-\rho(\tau, \varphi)\Delta\tau\Delta\varphi] \tag{6.6}$$

where $\rho(\tau, \varphi)$ is the joint distribution of scatterers with respect to delay time and Doppler shift φ, and $\rho(N, \tau, \varphi)$ is the probability density function of N(t), which represents the number of scatterers distributed in the range between $\tau \sim \tau + \Delta\tau$ and $\phi \sim \phi + \Delta\phi$.

We have assumed that the scatterers are independent of each other, and that $A_i(t)$ obeys the same distribution for all i. A_i, ϕ_i and t_i are independent of each other. The former indicates that the channel is frequency stable, and the latter indicates that the scattering with different Doppler frequency shifts is uncorrelated. Therefore, except that the propagation loss factor contained in $A_i(t)$ is a regular function of time t, this reverberation belongs to WSSUS channel.

Calculate the average of the system according to Eq. (6.5):

$$R_{wr}(t, t') = R_A(t, t') \iint \rho(\tau, \varphi)e^{j2\pi\varphi(t-t')}z(t - \tau)z^*(t' - \tau)d\tau d\varphi \tag{6.7}$$

where $R_A(t, t') = \langle A(t)A(t')\rangle$, which is a slow variation and a regular function. Therefore, for the stabilized reverberation process:

$$R''_{wr}(\Delta t) = K \iint \rho(\tau, \varphi)e^{j2\pi\varphi\Delta t}z(t - \tau)z^*(t' - \tau)d\tau d\varphi$$

$$= K \iint \rho(\tau, \varphi)e^{j2\pi\varphi\Delta t}R_z(\Delta t - \tau)d\tau d\varphi \tag{6.8}$$

where K is constant factor during operation.

Comparing Eq. (6.8) with Eq. (4.45), we can see that the power response function $R''_{hr}(\tau, \Delta t)$ of reverberation channel is:

$$R''_{hr}(\tau, \Delta t) = K \int \rho(\tau, \varphi) e^{j2\pi\varphi\Delta t} d\varphi \tag{6.9}$$

Figure 4.9 shows that $R''_{hr}(\tau, \Delta t)$ and reverberation scattering function $R''_{sr}(\tau, \varphi)$ are Fourier transform each other. Therefore, it is noted from Eq. (6.9):

$$R''_{sr}(\tau, \varphi) = K\rho(\tau, \varphi) \tag{6.10}$$

Equation (6.10) shows that the form of reverberation scattering function is determined by the distribution function of τ and ϕ in scatterers. Generally speaking, it is a joint distribution function of τ and φ. It only exists when the spatial distribution and Doppler distribution of scatterers are independent of each other:

$$R''_{sr}(\tau, \varphi) = K\rho(\tau)\rho(\varphi) \tag{6.11}$$

If the distribution of scatterers by distance is statistically uniform, the reverberation scattering function is:

$$R''_{sr}(\tau, \varphi) = K'\rho(\varphi) = \frac{K}{T}\rho(\varphi) \tag{6.12}$$

where K' is the constant;
  T is the pulse width of signal.

2.   Reverberation spectrum

If the sound source emission signal is $z(t)$, the autocorrelation function of reverberation can be seen from Eq. (4.45):

$$
\begin{aligned}
R''_{wt}(\Delta t) &= \int R''_{hr}(\tau, \Delta t) R_z(\Delta t - \tau) d\tau \\
&= \int G(\Delta f, \Delta t) R''_{Hr}(\Delta f, \Delta t) e^{j2\pi\Delta f\Delta t} d(\Delta f)
\end{aligned}
\tag{6.13}
$$

where $R_z(\tau) = \int_0^T z(t)z^*(t+\tau)dt$; $G(\Delta f, \Delta t) = G(f)$ is power spectrum of transmitted signal.

According to Eq. (6.13), the power spectrum of reverberation is:

$$G''_r(f) = G(f) \cdot R''_{Hr}(\Delta f, \Delta t) \tag{6.14}$$

6.3 Doppler Spread of Reverberation Spectrum

According to the scattering function of reverberation, the time spread of reverberation is large and the spectrum spread of reverberation is small. Therefore, it is hopeful to use the Doppler effect of moving target echo to detect the signal from the reverberation background. For the stabilized reverberation, $R''_{sr}(\varphi)$ can be measured with a long CW pulse. In the following description, "''" is omitted for convenience. The scattering function of the WSSUS component of the stabilized reverberation is recorded as $R_{sr}(\varphi)$. The principle of measuring $R_{sr}(\varphi)$ is described below.

According to Eq. (4.47), the output covariance $E(\tau_0, \varphi_0)$ of the matched filter in the reverberation channel is:

$$E(\tau_0, \varphi_0) = \iint R_{sr}(\tau, \varphi) |\chi(\tau_0 - \tau, \varphi_0 - \varphi)|^2 d\tau d\varphi$$

$$= \iint R_{sr}(\varphi) |\chi(\tau_0 - \tau, \varphi_0 - \varphi)|^2 d\tau d\varphi \qquad (6.15)$$

If the transmitted signal is a CW pulse with a sufficient pulse width, the Doppler cross section of its ambiguity function is $\delta(\varphi_0 - \varphi)$, which can be obtained by substituting formula (6.15):

$$E(\tau_0, \varphi_0) = R_{sr}(\varphi_0) \int |\chi(\tau_0 - \tau, 0)|^2 d\tau \qquad (6.16)$$

Equation (6.16) indicates that the output of the matched filter tuned at different frequencies is the form of the reverberation scattering function $R_{sr}(\varphi)$. For long CW pulse, the narrow-band filter with bandwidth of $1/T$ is an excellent approximation of the matched filter, so the spectrum analyzer can be used to measure $R_{sr}(\varphi)$.

Figures 6.8, 6.9, 6.10 and 6.11 show the experimental results of $R_{sr}(\varphi)$. CW pulse with acoustic frequency of 7–8 kHz and width of 1 s is used in the experiment, that is, the Doppler resolution of the signal is 1 Hz, and the sea state is level 1–2 during the experiment. The stabilization technology is used in the experiment.

Figures 6.8 and 6.9 show typical reverberation pulses after spectrum analysis. The frequency resolution of the experiment shown in Fig. 6.8 is 1 Hz, and the interval of frequency marks is 5.6 Hz, which is equivalent to the Doppler frequency offset of 1 kn speed. The signal on the right side of the figure is the target echo, and the target approaches the receiving point at 4 kn speed. The signal on the left is reverberation, and the interval between the two pulses is 8 s. The front part of the reverberation is sea surface reverberation. Due to the motion of sea surface waves, the Doppler spread is large, and the rear part of the reverberation is sea bed reverberation, with almost no Doppler spread. Figure 6.9 shows the time-varying spectrum of reverberation. 500 Hz is equivalent to zero Doppler (acoustic center frequency is 8 kHz), and the Doppler of 0.75 kn is corresponding to the Doppler of 4 Hz.

The experimental results of R. S. Thomas [3] are shown in Fig. 6.12. It is obvious that the Doppler spread of sea surface reverberation is asymmetric. Compared with

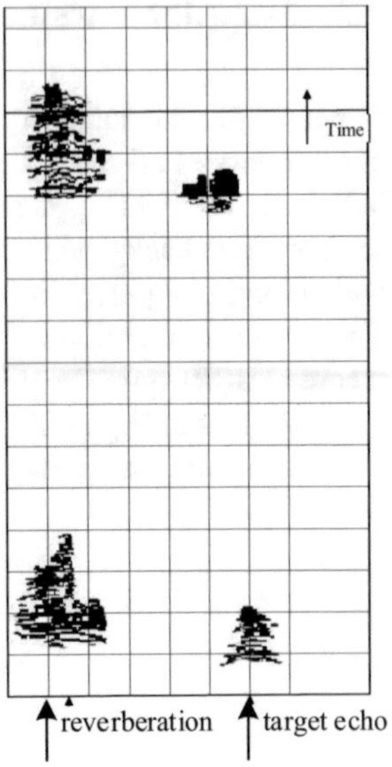

Fig. 6.8 The reverberation
and echo displayed on
optical (τ, φ) pseudo 3D
screen

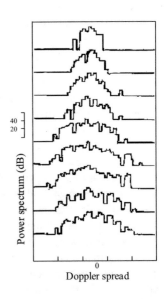

Fig. 6.9 The normalized
(smoothed) reverberation
Doppler spread

Fig. 6.12 (a) and (b), it is obvious that there is no obvious Doppler spread of seabed
reverberation.

Figure 6.10 shows the Doppler spread at different wind speeds. Doppler spread
limits the gain of Doppler processing and the accuracy of measuring target velocity.
The experimental results show that the target with relative velocity greater than $1\frac{3}{4}$
kn is easy to be detected when detecting the target with intensity above -6 dB under

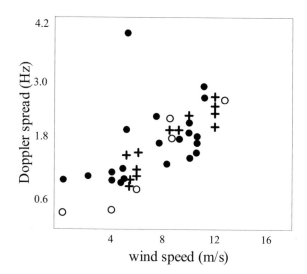

Fig. 6.10 The relationship between Doppler spread of sea surface reverberation and wind speed. •— depth <300 m, +—surface channel the results of R. S. Thomas (the results of R. S. Thomas). O—depth is 600 m (the results of J. Johnsen)

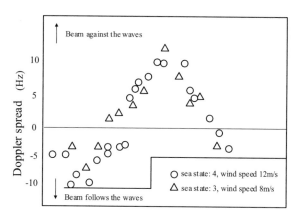

Fig. 6.11 Doppler shift of sea surface reverberation when the ship makes a steady circular motion in 20 min

level 4 sea state. Figure 6.11 illustrates the influence of sound wave transmission direction and wave angle on Doppler frequency shift.

6.4   Other Statistical Characteristics of Reverberation

Reverberation is formed by the superposition of backscattered waves from many scatterers at the receiving and transmitting points at the same time, and each scattered wave has little contribution to reverberation. Therefore, according to the law of large numbers, the instantaneous value of reverberation should meet the Gaussian distribution, and its envelope distribution should meet the Rayleigh distribution law. Experiments show that these conclusions are basically correct, especially for volume

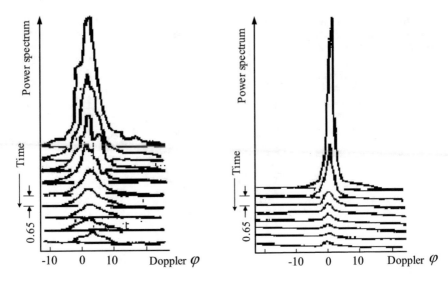

Fig. 6.12 The experimental results of R. S. Thomas (the frequency of long CW pulse is 9.5 kHz). **a** Sea surface reverberation. **b** Seabed reverberation

reverberation, but for seabed reverberation, especially for rock seabed, there are individual strong scattering points, which often deviate significantly from Gaussian distribution and Rayleigh distribution. By analyzing Eq. (6.13), it can be pointed out that the autocorrelation function of reverberation instantaneous value is very close to the signal autocorrelation function. The time correlation radius of the reverberation envelope is inversely proportional to the bandwidth of the transmitted signal. For CW pulses, the reverberation envelope is correlated within the pulse width. Literature [1] points out that the autocorrelation function of the reverberation envelope is:

$$R_A(\tau) \approx (1 - \tau/T)^2, \tau \leq T \tag{6.17}$$

where T means CW pulse width.

The offshore test results are reasonably consistent with the theory, as shown in Fig. 6.13 [2]. The solid line in the figure is the theoretical result.

The solid line in Fig. 6.14 is the theoretical relationship curve between the time autocorrelation radius of LFM signal envelope and TW (product of bandwidth and pulse width), and the ordinate is the normalized correlation time radius. The experiment shows that when the TW product is 6–8, the correlation of the echo envelope of the point target is significantly stronger than that of the reverberation. The experiment shows that the signal-to-mixing ratio gain of 8 dB may be obtained by using the post detection integral processing. Some sink detection sonars use the above principle.

The reverberation channel is also a random space-varying channel. Assuming that the channel is also generalized stationary in space, the channel can be described by spatial correlation function, which is defined as:

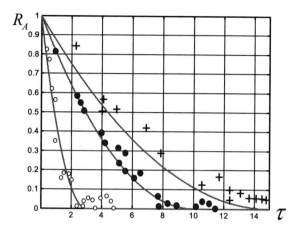

Fig. 6.13 The autocorrelation function of CW pulse envelope (1—T = 3 ms; 2—T = 10 ms; 3—T = 15 ms)

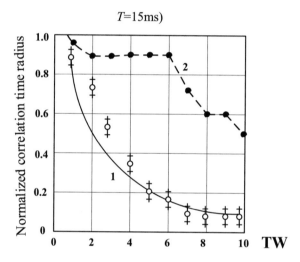

Fig. 6.14 The comparison of correlation time radius between echo envelope and reverberation envelope of FM signal point target (1—reverberation; 2—target)

$$R_{Hr}(r - r') = \langle H_r(f, r) H_r^*(f, r') \rangle$$

where r, r' is the radial vector of the spatial coordinates of the two receiving points. Here we will not discuss the relevant theories related to reverberation space, but only introduce some experimental data.

Figure 6.15 shows the transverse correlation characteristics in the sea surface reverberation horizontal plane. r is the distance between two receiving points, the ordinate is the transverse spatial correlation coefficient, and the abscissa is the normalized hydrophone spacing coordinate. It can be seen from the figure that the horizontal transverse correlation radius is about 4 times the wavelength. The spatial correlation radius is of great significance for array design and estimation of signal mixing ratio gain.

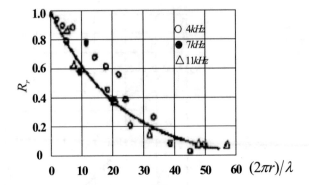

Fig. 6.15 The spatial correlation function of sea surface reverberation (related to transverse space in horizontal plane)

Table 6.1 The correlation between various reverberation successive pulses

Reverberation type	Variance (dB)			Correlation coefficient R ($T = 48$ s)		
	1.6 (km)	4 (km)	8 (km)	1.6 (km)	4 (km)	8 (km)
Volume	1.0	0.5	0.3	0.3	0.0	0.1
Sea surface	2.6	2.2	1.5	0.4	0.6	0.5
Seabed	3.2	3.9	3.3	0.7	0.8	0.7

The coherence of seabed reverberation is often strong. Clay and Medwin [3] studied the correlation between successive pulse reverberation, that is, the correlation between the reverberation envelope of one transmitted signal and the next transmitted signal. The CW pulse width used by him is 8 ms and the working frequency is 9.5 kHz. The results show that the envelope correlation of the two reverberation samples separated by 48 s is very strong, and the correlation coefficient is about 0.8. If we consider the translation of the transducer to the seabed during the experiment, the correlation coefficient may be 0.2 greater than that listed in Table 6.1 when the transducer is stationary.

6.5 Anti-reverberation

Reverberation is an important interference of active sonar. How to deal with reverberation in underwater acoustic equipment?

Different underwater acoustic equipment have different treatment methods for reverberation due to different service conditions and purposes. This section will give an overview.

The narrower the beam, the smaller the reverberation volume or area, so the smaller the reverberation energy will be received by the array. Therefore, the array beam width should be as narrow as possible while ensuring the requirements of other tactical use. For a particularly narrow beam, the side lobe of the beam should be carefully designed to be low enough to ensure that the reverberation received by the main lobe is greater than the total reverberation received by the side lobe. When the side lobe is large, narrowing the main lobe will not significantly improve the signal-to-reverberation ratio.

The reverberation must be stabilized or normalized before entering the processor. Generally, the normalization process should ensure that the output change does not exceed 1 dB when the input signal changes within 70 dB (or even 120 dB). The variation of the output of normalization processing is called normalization accuracy, which will affect the detection effect. Time gain control, automatic gain control and logarithmic amplifier are commonly used normalization techniques. The special discussion of normalization techniques is beyond the scope of this book.

When detecting low-speed targets in reverberation background, the most suitable waveform is still long CW pulse. In the ideal channel, when detecting the signal in the reverberation background, the ambiguity function of the best waveform is the thumbtack, but in the underwater acoustic channel, the joint ambiguity function between it and the channel is not thumbtack, and there are often many messy and high sidelobes. Therefore, the detection performance of the thumbtack function waveform is poor, unless the channel information can be extracted in real time and processed adaptively.

When using long CW pulses, the Doppler spread $R_{sr}(\varphi)$ of reverberation determines the measurement accuracy of target velocity and limits its detection performance. Experiments show that it is difficult to detect moving targets with $1\frac{3}{4}$ kn velocity lower than -6 dB under level 4 sea conditions and limited sea surface reverberation.

References

1. Olishevskii VV. Statistical properties of marine reverberation; 1966.
2. Thomas. Proceedings of the advanced symposium on sonar signal processing; 1976.
3. Clay CS, Medwin H. Acoustical oceanography principles & applications; 1977.

Chapter 7
Active Sonar Target Channel

The active sonar system sends out a high-intensity detection acoustic pulse. The ocean channel transmits the acoustic signal to the target. The echo (scattered sound wave) of the target reaches the receiving array again through the ocean. The sonar detects and analyzes the target information carried by the echo, in order to decide whether there is a target and what the types the target has. In the sonar equation, only the target strength is used to describe the characteristics of the target. In order to study the target recognition technology of active sonar, the description of the target characteristics must be deepened. In this chapter, the target is regarded as "multiple highlight target channel".

The characteristics of target serves for detection, recognition, and target parameter estimation. With the sonar moving farther and farther away, the importance of active sonar to target recognition has attracted more and more attention. When a long-range sonar detects many targets in its field of vision, it must distinguish the interested targets in order to improve its combat efficiency. Torpedo acoustic homing system is a kind of short-range sonar, which is in the near field of the target. The modern torpedo can distinguish the authenticity of the target, that is, the acoustic homing system is required to distinguish the real target and the acoustic decoy in the medium and short range (300–800 m) to ensure the attack effect of the torpedo.

The incident sound wave is projected onto the target to produce scattered sound wave. From the point of view of communication theory, the target makes a linear transformation of the incident sound wave and transforms the intensity and waveform of the incident sound wave. Therefore, it can be regarded as a time–space filter, which is called "target channel". The target channel is a link in the information transmission chain of active sonar. The information flow of active sonar is shown in Fig. 7.1.

The target echo consists of two parts: coherent component and random component. The former is a strong "geometric mirror reflection" component, and the latter is the scattering component generated by the edge, angle and wake of the target. Figure 7.2 shows that the target channel is composed of two types of filters, which correspond to the above two components respectively.

© Harbin Engineering University Press 2022
J. Hui and X. Sheng, *Underwater Acoustic Channel*,
https://doi.org/10.1007/978-981-19-0774-6_7

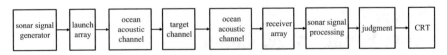

Fig. 7.1 The schematic diagram of active sonar information flow

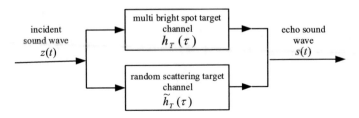

Fig. 7.2 The schematic diagram of target channel

7.1 Point Targets

A real target is by no means a point. The sound wave incident on the target generates scattered sound wave. If the backscattering can is radiated by its equivalent acoustic center, and the equivalent acoustic center is a point, it is called point target or point target model.

In the sonar equation, the point target intensity is defined as the decibel of the ratio of echo (backscattered wave) intensity to incident sound intensity at 1 m from the target acoustic center. The point target strength T_s is:

$$T_s = 10 \log \frac{I_r}{I_i}\bigg|_{r=1m} \tag{7.1}$$

where I_r and I_i are the echo sound intensity and incident wave sound intensity "1 m" away from the acoustic center, respectively.

The so-called "acoustic center" is an imaginary reflection point that may be inside, outside or on the boundary of the target. From a distance, the echo is radiated from this point.

When "1 m" the reference distance, the target intensity of many sonar targets is positive, which does not mean that the target has focusing effect when scattering and it makes the echo stronger than the incident sound. If "1 km" is taken as the reference distance, the values of almost all targets will be negative.

The target intensity of simple shape targets can be calculated by wave acoustics and ray acoustics, and the target intensity and scattering characteristics of complex targets can also be predicted by finite element and boundary element methods.

Next, the target strength of a rigid ball with a diameter larger than the wave will be calculated by the ray acoustic method. The calculation process can further explain the meaning of the target strength.

If the plane wave with sound intensity I_i irradiates on a rigid ball with radius a, the incident sound power intercepted by the ball is $\pi a^2 I_i$. Because the ball is rigid, it radiates this power uniformly in all directions according to the law of spherical wave again, and the echo intensity I_r from the ball center r is:

$$I_r = \frac{\pi a^2 I_i}{4\pi r^2} = I_i \frac{a^2}{4r^2}$$

Take $r = 1$ m, and it becomes:

$$TS = 10\log\frac{a^2}{4} \tag{7.2}$$

According to the above equation, the target intensity of a rigid ball with a radius of 2 m is just 0 dB.

The far-field point target intensity of simple target obtained by wave acoustic and ray acoustic methods is listed in Table 7.1.

Generally speaking, the target strength is also related to the incident acoustic frequency. Equation (7.2) is the target strength of the large steel ball ($a \gg \lambda$). The

Table 7.1 Target strength T_s of rigid object with simple shape $TS = 10\log A$, r is the distance, k is the wave number

Target shape		Target strength $<T_s = 10\log A>$ A	Symbol definition	Incident direction	Condition
Arbitrary shape surface		$\frac{a_1 a_2}{4}$	a_1, a_2—orthogonal principal radius of curvature	Perpendicular to surface	$ka_1 \gg 1, ka_2 \gg 1$ $r > a_1, r > a_2$
Sphere	Large	$a^2/4$	a—sphere radius	Random	$ka \gg 1, r > a$
		$61.7\frac{V^2}{\lambda^4}$	V—ball volume λ—wavelength	Random	$ka \ll 1$ $kr \gg 1$
Infinite cylinder	Crude	$ar/2$	a—column radius	Perpendicular to column axis	$ka \gg 1$ $r > a$
	Fine	$9\pi^4 a^4/\lambda^2$	a—column radius	Perpendicular to column axis	$ka \ll 1, r > a$
Finite length cylinder		$aL^2/2\lambda$	L—column length a—column radius	Perpendicular to column axis	$ka \gg 1$ $r > L^2/\lambda$
		$\left(\frac{aL^2}{2\lambda}\right)\left(\frac{\sin\beta}{\beta}\right)^2\cos^2\theta$	a—column radius $\beta = kL\sin\theta$	θ angle to normal	$ka \gg 1$ $r > L^2/\lambda$

Fig. 7.3 General relationship between steel ball target strength and acoustic frequency. (The ordinate a_e^2/a^2 is the square of the ratio of the equivalent scattering sphere radius to the real sphere radius)

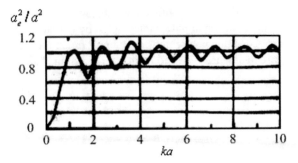

Fig. 7.4 The relationship between near-field target strength and distance of rigid disc

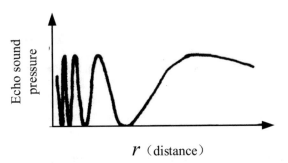

general relationship between the steel ball target strength and the acoustic frequency is shown in Fig. 7.3. When $ka < 0.5$, the target strength of steel ball increases with the fourth power of frequency.

For large target, the target intensity in the near-field region oscillates with the distance. This is due to the interference effect of the scattered sound near-field, which leads to different "Brightness" (target intensity) of the target observed at different distances. When the observer gradually approaches the target, it observes that the brightness of the target is flickering. If the incident acoustic wave is not a CW signal but a limited bandwidth acoustic signal, the flicker of the target brightness will be reduced, which is the comprehensive effect of many frequency components of the limited bandwidth signal. As an example, the flicker of short-range target intensity when using long CW pulse acoustic wave is shown in Fig. 7.4.

7.2 Coherent Point Target Channel

In the "multiple highlight point target" model of a real target, each coherent highlight is a point target and is associated with a point target channel. In Fig. 7.2, the impulse response of the coherent component of the point target channel is $h_T(\tau)$, and its Fourier transform is the frequency response $H_T(f)$ of the point target channel. In infinite space, the target echo $H_T(f)$ is:

$$s(t) = \frac{1}{r}z(t) * h_T(t, r) \tag{7.3}$$

$$S(f) = \frac{1}{r}Z(f) \cdot H_T(f, r) \tag{7.4}$$

where, r is the distance, $z(t)$ is the incident waveform and $Z(f)$ is its spectrum.

It can be seen from Fig. 7.2 that the channel frequency response function $H_T(f)$ of the spherical target is the frequency response function of the spherical target in the far field. When the frequency is high ($ka \gg 1$), the frequency response is "white", i.e. independent of frequency; When the frequency is low ($ka < 1$), the target frequency response of the sphere is equivalent to a high pass filter, and its frequency response amplitude is proportional to the frequency to the power of four, which is called Rayleigh scattering.

It can be seen from Fig. 7.4 that the target impulse response and frequency response are related to the distance in the near field, which are $h_T(t, r)$ and $H_T(f, r)$. The "Brightness" of the target is "flashing" with the distance. For broadband signals, the average intensity of the target is stable in the total bandwidth. Therefore, the echo intensity of broadband sonar is relatively stable, and the fluctuation of CW pulse echo intensity is larger.

7.3   Multiple Highlight Target Model

Submarine is the most interesting target of sonar and torpedo homing, and it is regarded as a multiple highlight target model.

Submarine has complex interface shape and internal structure, and its scattering characteristics are very complex. According to the experimental results, a multiple highlight target model is established as follows:

- Geometric mirror reflection highlight of the hull: when the normal of the local surface of the hull is parallel to the incident sound axis, it becomes the geometric mirror reflection highlight. It is a moving highlight with high target intensity, that is, when the incident side angle is different, highlights are formed in different parts of the hull.
- The geometric mirror reflection highlight of the hull of the boat bridge is referred to as the hull highlight for short: Although the hull highlight appears in different parts according to the incident direction on the hull, it is essentially a fixed highlight because the hull scale is only a few meters. The echo intensity of highlights in the enclosure is also large.
- Special highlights of water tank: when the submarine floats or dives, the lifting water tank forms special highlights with high echo intensity within some incident side angle range. This is a strong highlight with a fixed position, but it only appears in a specific angle range. Special highlights will significantly expand the echo length.

- Propeller highlight: This is a kind of fixed highlight with medium echo intensity.
- Wake scattering echo: only surface ships and submarines sailing on the surface can produce wake echo. For the high-frequency torpedo acoustic homing band, the wake echo is very strong, and the target scale of the wake echo is also very large.
- Edge and angle scattering echo: the submarine has a complex structure, and its ribs, pipes and other structural parts have many edges and angles. They produce weak random scattering, which is distributed in the scale range of the whole boat, so they are called "highlight background".

Figure 7.5 shows the sea trial results of submarine multi-bright-spot echo. The beam width of the measurement system is 7°, the signal for measurement is CW narrow pulse, and the pulse width is 1 ms. The target distance is 200 m. "1" in the figure refers to three transponders installed on both sides of the hull of the boat bridge, with three white symbols respectively " ×", "+", "△". Their positions are determined by precision direction finding and response ranging. Therefore, the highlight (black) sandwiched between the three transponders must be the highlight of the enclosure. "2" indicates the four main highlights of the boat, which are "geometric mirror reflection highlight of the boat shell", "enclosure highlight", "special highlight of the water tank" and "propeller highlight" from bottom to top. The abscissa of the figure is azimuth and the ordinate is time delay (i.e. distance). "3" is the multi-path reflection of highlight echo, and a series of reflected sound arrive in the figure. In the azimuth time history range covered by the main highlight, the brightness (i.e. echo intensity) is also high, which is called "highlight background". They are scattered echoes generated by edges and corners "5" is the side lobe stage of the receiving array beam and also has a large output power stage.

The sea trial results in Fig. 7.5 fully prove the rationality of the "multiple highlight target model".

Figure 7.5 shows that at about 200 m, the azimuth width occupied by multiple highlights is about 30°, which is referred to as "azimuth width of the target", also known as "transverse scale" of the target. The farther the target is, the smaller the azimuth width of the target must be. It can be inferred from Fig. 7.4 that the azimuth width of the target at 800 m is only a few degrees. The single transponder acoustic decoy is a point target, and its "azimuth width" is very small, while the azimuth width

Fig. 7.5 The sea trial results of multiple highlight echo of submarine target

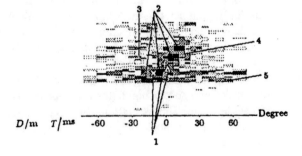

Fig. 7.6 The multiple
highlights sea trial results of
submarine periscope during
deep navigation

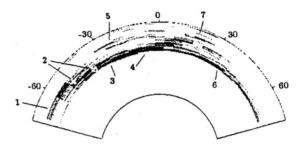

of submarine target at 800 m is much larger than that of point transponder. This is
the basic principle of scale recognition of torpedo acoustic homing target.

Due to the small size of torpedo, the size of acoustic homing array is small, and
the beam width of array is about 20°. Such a wide beam cannot be used to estimate
the "lateral scale" of the target (i.e. the azimuth width occupied by the target). The
split array phase difference estimation technique is needed to measure the lateral
scale of the target. Technical details are beyond the scope of this tutorial, so we have
to omit it here.

Figure 7.6 is the multiple highlights echo diagram of submarine periscope during
navigation. The figure is a polar graph, and the direction of the radiation line is
distance (also means time delay).

"1" in Fig. 7.6 represents the echo highlight generated by the bow surge, and
"2" is the measured position of two transponders, indicated by "×" in white. They
are respectively located on the front and rear sides of the bridge enclosure, and
the highlight between them is the enclosure highlight. "3" and "4" are the main
highlights. "5" is the multi-path reflection of highlight echo. "6" is the strong side
lobe stage of the output. The scale of the target highlight set is significantly larger
than that of the captain, because the wake scattering significantly increases the scale
of the highlight set. The high brightness band between "4" and "7" is the broadband
of the wake.

Ships sailing on the surface have long wake zone, which is the difference between
surface targets and "multiple highlight (multi transponder) acoustic bait".

It can be seen from the discussion in this section that the echo characteristics of
the target are determined by the system function $h_T(t, r, \alpha)$ of the target channel. α
is the side angle of the incident sound beam axis relative to the boat. If the side angle
α of the incident sound is different, the multiple highlight structure of the target is
different. In essence, the fundamental of active acoustic homing target recognition
is to estimate the impulse response function $\hat{h}_T(t, r, \alpha)$ of the target channel. Target
lateral scale recognition technology is only a preliminary azimuth domain estimation
of all target characteristics.

At a close distance of 200 m, the multi beam system with narrow beam can
separate the echoes of different highlights in azimuth domain. Figure 7.7 shows the
sea trial results of CW narrow pulse echo. Figure 7.7a shows the echo waveform of
the highlight reflected by the geometric mirror, which has steep rising and falling

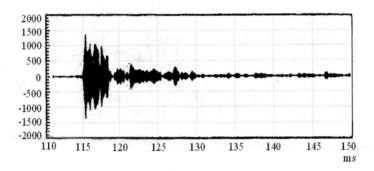

(a) reflection highlight echo waveform of narrow CW pulse geometric mirror

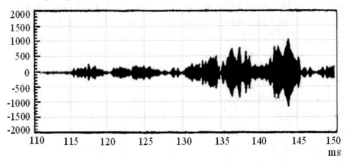

(b) special highlight echo waveform

Fig. 7.7 The echo waveforms of two different highlights

edges. Due to the multi-path reflection and interference of the ocean interface, the echo is widened by about 3MS. Figure 7.7b shows the echo waveform of the special highlight of the water tank. It is a pulse train echo, a series of echo pulses with gradually increasing amplitude, and the echo is widened for tens of milliseconds. This shows two highlights with different properties. They have impulse response functions with very different properties. Estimating the impulse response of the target can distinguish the target more effectively.

7.4 General Characteristics of Submarine Target Strength

In the previous section, in order to study the multiple highlight target characteristics of submarines, narrow CW pulses are used. When the distance is very close, the echoes of different highlights can be distinguished in azimuth and time. Sonar uses long pulse to detect the target at a long distance. At this time, the echoes of each highlight are superimposed. The target intensity formed by interference and superposition of multiple highlight echoes is directional.

The target strength is related to the incident side angle, which is called the directivity of the target strength. The typical characteristics are shown in Fig. 7.8.

There are only two important highlights when incident in the bow and stern directions. They are the highlights reflected by the geometric mirror of the bow and enclosure (when incident in the bow direction); Highlights of propeller and enclosure (when incident in the stern direction). The number of highlights contributing to the echo is small, and the intensity of each highlight is small, so the target intensity in the head and tail directions is less than 10 dB. In the case of normal transverse incidence, there are many important highlights, and the intensity of each highlight is large, so the target intensity value in the normal transverse direction is large, which can reach 25 dB.

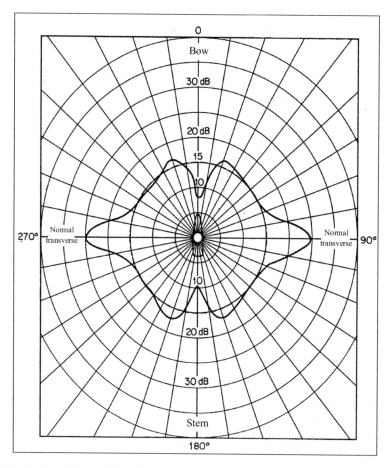

Fig. 7.8 The butterfly directivity of submarine target strength

Because the target and the sonar platform move relatively, there is a Doppler frequency offset between the frequency of the echo and the sonar transmitted signal frequency. Sonar signal processing must have enough Doppler tolerance and adopt Doppler compensation signal processing technology.

Due to the multi-path interference effect of ocean channel and the interference of multiple highlight echo of target, the CW pulse echo intensity fluctuates, usually its fluctuation will exceed 10 dB, which makes the detection effect unstable. The interference effect is blurred by multi frequency signal or broadband signal, and their echo intensity is relatively stable.

Chapter 8
Coherent Structure of Shallow Acoustic Field

Chapters 2 and 3 introduce the coherent characteristics of ocean acoustic channel, and in Chaps. 4 and 5, we have discussed the incoherent characteristics of channel. For one-way propagation, 95% of the energy is coherent, and for high-frequency reverberation, incoherent energy is the main element.

The working frequency of modern sonar is developing towards low quantity. The shallow water sound field below 1 kHz has a stable interference structure. Even for reverberation, it is a random process, but reverberation below hundreds of Hz also has a stable interference structure.

The coherent structure of shallow water channel and its application are the research hotspots of modern underwater acoustics. It can be used to estimate the motion parameters of targets and improve the ability of underwater acoustic system to detect and recognize targets.

This chapter will introduce the basic knowledge of coherent structure of shallow water sound field and its application.

8.1 Coherent Structure of Short-Range Sound Field in Shallow Water [1]

This section will introduce the coherent structure of continuous spectral sound field of a moving sound source in shallow water and the estimation of target motion parameters.

If a sound source (target) radiating continuous spectrum noise moves in a straight line at Ulanhot speed at equal depth in shallow water, as shown in Fig. 8.1. It marks that the horizontal distance between it and the hydrophone is $r(t)$ at the moment of t; The closest distance between the target and the hydrophone $r(t_0) = r_0$ at the moment of t_0.

© Harbin Engineering University Press 2022
J. Hui and X. Sheng, *Underwater Acoustic Channel*,
https://doi.org/10.1007/978-981-19-0774-6_8

Fig. 8.1 The geometry of movement

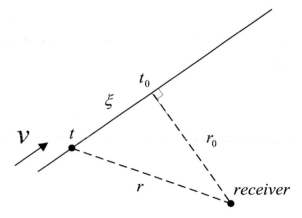

There are many descriptions of sound field coherence, and one of them is the coherent fringe on the LOFAR diagram, as shown in Fig. 8.2. The continuous spectrum noise is radiated during the movement of the target, and the short-term power spectrum analysis is made for the received signal of the hydrophone. The time history of the time series of the short-term power spectrum is LOFAR diagram, as shown in Fig. 8.2. Figure 8.2 is the LOFAR diagram of Baa. For a given timing, it marks that the frequency slice in the diagram is the power spectrum of the received signal at that time (shown in formula B). The figure shows that there are obvious interference fringes in the sound field. This section will explain that the interference fringes in the figure are a series of hyperbolas, and the parameters of the hyperbola can be extracted to estimate the motion elements of the target.

The coherent component of shallow water sound field can be expressed as Eq. (3.48). The first and second terms of this formula constitute a sea surface dipole, as shown in Fig. 8.3; The first and third items constitute the seabed dipole, as shown

Fig. 8.2 Interference fringes of short-range sound field

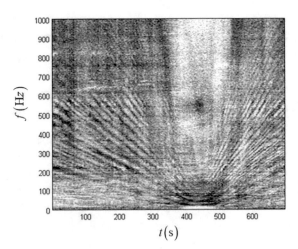

Fig. 8.3 The hyperbolic fringe of sea surface dipole

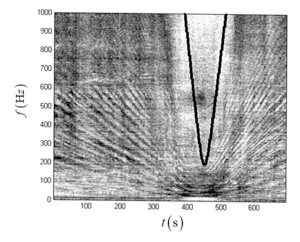

in Fig. 8.4. In other words, the direct sound and the reflected sound form a sea surface dipole; The direct sound and the reflected sound form a seabed dipole. For a better understanding, note that at the point source $(0, z_1)$, it radiates the harmonic sound wave $(0, z_s)$, and it radiates the harmonic sound wave $e^{j\omega t}$. Then the sound pressure (r, z) received at the receiving point RZ is

$$p_s(r, f) = \frac{1}{R_1}e^{-jkR_1} - \frac{1}{R_2}e^{-jkR_2} \tag{8.1}$$

n the above formula, the first term is direct sound, and the second term is sea surface reflected sound, in which the sea surface reflection coefficient is -1. Both of the two terms constitute the sea surface dipole sound field. Where, $k = \frac{\omega}{c}$, c is the sound

Fig. 8.4 The hyperbolic fringe of seabed dipole

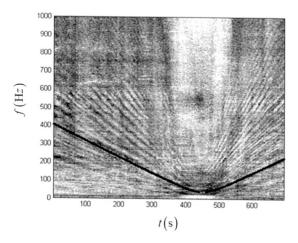

velocity of sea water. And

$$R_1 = \sqrt{r^2 + (z - z_1)^2} \tag{8.2}$$

$$R_2 = \sqrt{r^2 + (z + z_1)^2} \tag{8.3}$$

If the skew distance is $R_1 \gg z$, $(i = 1, 2)$, Eq. (8.1) can be approximately

$$p_s(r, f) \approx -\frac{2j}{R_1} e^{jkR_1} \sin \frac{kz_s}{R_1} \tag{8.4}$$

The sinusoidal function in the above formula represents the sea surface dipole interference fringes. When Eq. (8.4) is zero, i.e., when

$$\frac{kz_s z}{R_1} = l\pi \quad (l = 0, 1, 2, \ldots)$$

The above formula determines the dark fringe of the sea surface dipole in LOFAR diagram. The above formula appears to be

$$\frac{kz_s z}{\sqrt{r^2 + (z - z_s)^2}} = l\pi \quad (l = 0, 1, 2, \ldots) \tag{8.5}$$

Let

$$\tau = t - t_0$$

So

$$r = \sqrt{r_0^2 + (\upsilon\tau)^2} \tag{8.6}$$

where υ is the target speed. Bring Eq. (8.6) into Eq. (8.5). After sorting, the equation of the dark interference fringe in the LOFAR diagram is

$$\frac{f^2}{a_s^2} - \frac{\tau^2}{b_s^2} = 1 \tag{8.7}$$

where

$$a_s = \frac{lc\sqrt{r_0^2 + z^2}}{2z_s z}, b_s = \frac{\sqrt{r_0^2 + z^2}}{\upsilon} \tag{8.8}$$

Fig. 8.5 Sea surface dipole LOFAR interference fringes

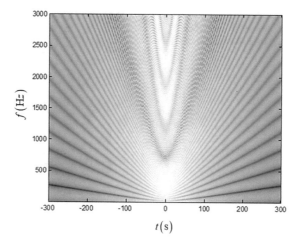

Equation (8.7) is the standard form of hyperbolic equation. Therefore, in the LOFAR diagram, the dark fringes of Shanghai dipole interference are a group of hyperbolas, which have the same wedge center, i.e. $\tau = 0$.

The structure diagram of sea surface dipole interference is simulated and calculated. It is assumed that the target moving speed is $v = 5\,\mathrm{m/s}$, the nearest distance through the hydrophone is $r_0 = 128\,\mathrm{m}$, the target depth $Z_s = 3.6\,\mathrm{m}$, the hydrophone depth $Z = 45.5\,\mathrm{m}$, and the radiated noise frequency band is 1–3 kHz. At this time, the sea surface dipole dark fringe given by the simulation is shown in Fig. 8.5.

Similarly, the seabed dipole sound pressure composed of direct sound and seabed reflected sound is

$$p_b(r, f) \approx -\frac{2jv_b}{R_2} e^{jkR_2} \sin\left(\frac{k(2H - z_s)z}{R_2}\right) \tag{8.9}$$

where v_b is the seabed reflection coefficient, $k = \frac{\omega}{c}$ is the beam, H is the sea depth, and

$$R_2 = \sqrt{r^2 + (2H - z_s)^2} \tag{8.10}$$

The trajectory equation of dark stripe of seabed dipole on LOFAR diagram is that Eq. (8.9) is equal to zero, i.e.

$$\frac{r^2 + (2H - z_s)z}{R_2} = l\pi, l = 1, 2, \ldots \tag{8.11}$$

Replace Formula (8.10) and Formula (8.6) into the above formula, and get it after sorting

Fig. 8.6 Interference conditions of sea surface dipole and seabed dipole

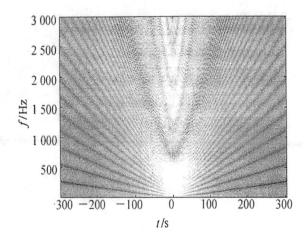

$$\frac{f^2}{a_b^2} - \frac{\tau^2}{b_b^2} = 1 \tag{8.12}$$

where

$$a_b = \frac{lc\sqrt{r_0^2 + (2H - z_s)^2}}{2(2H - z_s)z}, \quad b_b = \frac{\sqrt{r_0^2 + (2H - z_s)^2}}{v} \tag{8.13}$$

Equation (8.12) points out that the dark stripes generated by the bottom dipole in the LOFAR diagram are also a group of hyperbolas, and their vertex center coordinates are also $\tau = 0$. Under the same conditions in Fig. 8.5, when H = 47 m, the simulation results of double dipole stripes composed of sea surface dipole and seabed dipole are shown in Fig. 8.6.

According to the measured sound field interference structure, the target motion parameters can be estimated, such as target speed v, target depth z_s, target nearest passing distance t_0 and target distance $r(t)$. At this time, the sea depth H and the hydrophone depth z are a priori. The LOFAR diagram can be obtained by analyzing the power spectrum of the received signal. The hyperbolic parameter $a_s, b_s; a_b, b_b$ of the dark fringe is extracted according to the LOFAR diagram, and the motion parameters of the target can be estimated as

$$z_s = \frac{2Ha_bb_s}{a_bb_s + a_sb_b} \tag{8.14}$$

$$v = \frac{4Hza_ba_sb_b}{cb_b(b_ba_s + a_bb_s)} \tag{8.15}$$

$$r_0 = \sqrt{\left[\frac{4Hza_ba_sb_b}{cb_b(b_ba_s + a_bb_s)}\right]^2 b_s^2 - z^2} \tag{8.16}$$

The target skew distance R is

$$R = \sqrt{r_0^2 + v^2\tau^2 + (z - z_s)^2} \tag{8.17}$$

According to the LOFAR diagram measured in the experiment, the sea surface and seabed dipole parameters a_s, b_s; a_b, b_b can be obtained by Hough transform. The relevant detailed discussion is not introduced here, but can be referred to the relevant literature [2].

Figures 8.7 and 8.8 introduce the sea trial results. The thick black line in the figure is the dipole interference dark fringe extracted by Hough transform.

Fig. 8.7 Sea surface dipole fringe (thick black line) extracted from sea trial data

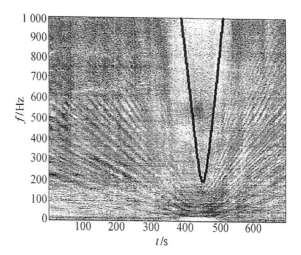

Fig. 8.8 Seabed dipole fringe (thick black line) extracted from sea trial data

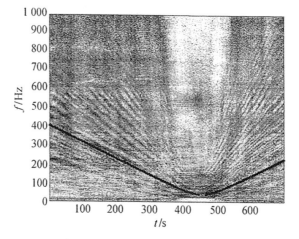

In Figs. 8.7 and 8.8, the parameters of dipole hyperbolic dark fringe extracted by Hough transform are $a_s = 192, b_s = 12; a_b = 37, b_b = 41$. Bring in Eqs. (8.14), (8.15) and (8.16) respectively to obtain target source depth $z_s = 5.8$ m, speed $v = 4.1$ m/s and nearest passing distance $r_0 = 109.2$ m. Compared with the true value, the errors are 3.3%, 8.9% and 16% respectively.

This section illustrates that the low-frequency sound field below 1 kHz has a stable interference structure, and LOFAR diagram is one of the description methods of sound field interference structure. On LOFAR diagram, the coherence of sound field is shown as interference fringes with light and dark, mainly sea dipole and seabed dipole interference fringes. The motion parameters of short-range targets can be estimated according to the hyperbolic parameters of interference dark fringes.

8.2 The Interference Structure of Normal Wave Sound Field in Shallow Water Waveguide

This section introduces the interference structure of vector sound field and its application. In the previous section, the interference fringes of LOFAR diagram are used to describe the interference structure of sound field. In this section, the positive and negative spatial distribution of vertical sound intensity flow is used to describe the interference structure of vector sound field.

8.2.1 The Sound Pressure Field of Pekeris Waveguide

In the middle of the twentieth century, the normal wave theory of point source sound field in shallow water layered media has been mature [3–7], but they only care about the sound pressure field, and few people pay attention to the particle vibration velocity field. The theory holds that the sound field is the sum of the normal wave and the side wave, which attenuates rapidly with the increase of distance. At a long distance, the normal wave of the waveguide (the part corresponding to the real eigenvalue) determines the sound field at the receiving point. The normal wave is a traveling wave in the horizontal direction and a standing wave in the vertical direction (see 2.8), that is, it is considered that sound energy is transmitted only in the horizontal direction.

Title sound field sound intensity $I = \langle p^2(t) \rangle / \rho c$, $\langle p^2(t) \rangle$ is the time average of the square of sound pressure, and ρc is the acoustic impedance of medium. The theory of sound propagation in layered media has promoted the rapid development of underwater acoustic technology since 1960s. When the theory is mature, vector sensor, vector acoustics and vector signal processing technology have just started, and then attracted the attention of domestic academic circles in the 1990s. Relative to the scalar field sound pressure p, the acoustics that pay attention to both sound pressure

Fig. 8.9 The schematic diagram of Pekeris waveguide

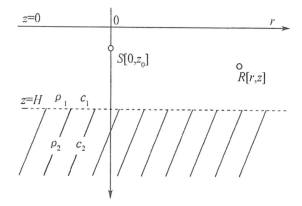

and particle vibration velocity (hereinafter referred to as vibration velocity) is called vector acoustics [8, 9]. Vector acoustics believes that the sound field characteristics (except the so-called simple sound field) must be described by sound too far and vibration velocity, and only p cannot fully describe the sound field characteristics. The fundamental definition of sound intensity is the average active sound intensity flow $\langle \mathrm{Re}\{pV^*\}\rangle$ per unit area. Here, $\mathrm{Re}\{\cdot\}$ represents the real part, and the superscript $*$ represents complex conjugate.

This section discusses the simplest model of sound propagation in shallow water, that is, the normal wave of point source sound field in Pekeris waveguide [7].

As shown in Fig. 8.9, the medium in the water layer is uniform, the density is ρ_1 and the sound velocity is c_1. In the cylindrical coordinate system, the sea level is $z = 0$ plane, the seabed is $z = H$ plane, and the positive direction of z axis points vertically downward. The point source is located at $[0, z_0]$, the receiving point is located at $[r, z]$, and r is the horizontal distance. The impedance of seabed medium is $\rho_2 c_2$; The sea surface is an absolute soft interface, and the sound pressure on it is zero. The point source radiates harmonic sound waves, the time factor is $\exp(-j\omega t)$, and ω is the angular frequency.

The sound field is the sum of normal wave and side wave, and the waveguide normal wave can be considered at a distance.

The particle beam potential function of normal wave in water layer is [7]

$$\varphi(r, z_0, z) = 2\pi j \sum_n \sin(\beta_{1n}, z) F(z_0, \xi_n) H_0^{(1)}(\xi_n, r), 0 \le z \le H \qquad (8.18)$$

The above formula omits the time factor and $H_0^{(1)}$ is the first kind of zero order Hankel function, where

$$F(z_0, \xi_n) = \frac{\beta_{1n} \sin(\beta_{1n}, z_0)}{\beta_{1n} H - \sin(\beta_{1n} H) \cos(\beta_{1n} H) - b^2 \tan(\beta_{1n} H) \sin^2(\beta_{1n} H)} \qquad (8.19)$$

And $\beta_m = \sqrt{k_i^2 - \xi_n^2}$, $b = \rho_1/\rho_2$, $k_i = \omega/c_i$ ($i = 1, 2$), n is the sequence number of the simple wave, ξ_n is the nth local value, which is the root of the following eigen equation

$$x \cos x - jb\sqrt{x^2 - \sigma^2} \sin x = 0 \tag{8.20}$$

where

$$x = \beta_1 H$$
$$\sigma^2 = (k_1^2 - k_2^2) H^2 \tag{8.21}$$

Each order normal wave corresponds to a cut-off frequency f_n, that is, when the acoustic frequency $f < f_n$, the nth order waveguide normal wave cannot be excited by the sound source. Here we have [7]

$$f_n = \frac{(n - \frac{1}{2}) c_1 c_2}{2H\sqrt{c_2^2 - c_2^2}} \tag{8.22}$$

8.2.2   The Normal Wave Sound Intensity Flow

The sound pressure p from Eq. (8.18) is

$$p(r, z_0, z) = 22\pi \omega \rho_1 \sum_n \sin(\beta_{1n} z) f(z_0, \xi_n) H_0^{(1)}(\xi_n r)$$

$$\approx e^{-j\frac{\pi}{4}} \sqrt{\frac{8\pi}{r}} \omega \rho_1 \sum_n \sqrt{\frac{1}{\xi_n}} \sin(\beta_{1n} z) f(z_0, \xi_n) e^{j\xi_n r} \tag{8.23}$$

In the above formula, the bulk asymptotic expansion of Hankel function is made. The relationship between vibration velocity V and sound pressure p is

$$\rho \frac{\partial V}{\partial t} = -\nabla p \tag{8.24}$$

Note that the time factor is $e^{-j\omega t}$, so the above formula is

$$V = \frac{1}{j\omega\rho} \nabla p = \frac{1}{j\omega\rho} \left[\frac{\partial p}{\partial r} \mathbf{i} + \frac{\partial p}{\partial z} \mathbf{k} \right] \tag{8.25}$$

Because the sound field is cylindrical symmetry, the term related to the horizontal azimuth in the above formula is zero. The derivative of Eq. (8.23) is derived from

Eq. (8.25), and the horizontal component of vibration velocity v_r is

$$v_r \approx e^{-j\frac{\pi}{4}} \sqrt{\frac{8\pi}{r}} \sum_n \sqrt{\xi_n} \sin(\beta_{1n}z) F(z_0, \xi_n) e^{j\xi_n r} \tag{8.26}$$

The vertical component v_z of vibration velocity is

$$v_z = -je^{-j\frac{\pi}{4}} \sqrt{\frac{8\pi}{r}} \sum_n \sqrt{\frac{1}{\xi_n}} \beta_{1n} \cos(\beta_{1n}z) F(z_0, \xi_n) e^{j\xi_n r} \tag{8.27}$$

Sound intensity water bisection is

$$V = p v_r^* \approx \frac{8\pi \omega \rho_1}{r} \sum_n \sin^2(\beta_{1n}z) F^2(z_0, \xi_n)$$

$$+ \frac{8\pi \omega \rho_1}{r} \sum_{n,n \neq m} \sum_m \sin(\beta_{1n}z) \sin(\beta_{1m}z) F^2(z_0, \xi_n) F(z_0, \xi_m)$$

$$\times \sqrt{\xi_m/\xi_n} \{\cos[(\xi_m - \xi_n)r] + j \sin[(\xi_m - \xi_n)r]\} \tag{8.28}$$

The first term on the right of the above formula is a real number, which is the horizontal sound intensity flow of each order normal wave, and it is the active sound intensity flow (real number). That is to say, for each normal wave alone, it handles the transport energy in the horizontal direction and is a traveling wave in the horizontal direction. However, the second term of Eq. (8.28) is complex, which shows that due to the mutual interference of multi-order normal waves, the horizontal sound intensity flow has both active and reactive components, indicating that even for the normal wave sound field, it also has the performance of vector sound field characteristics.

The active component I_{rA} and reactive component I_{rA} of horizontal sound intensity flow are

$$I_{rA} = \text{Re}(pv_r^*), \quad I_{rR} = \text{Im}(pv_r^*) \tag{8.29}$$

The vertical sound intensity flow is

$$I_s = pv_r^* \approx j\frac{8\pi \omega \rho_1}{r} \sum_n \frac{\beta_{1n}}{\xi_n} \sin(\beta_{1n}z) \cos(\beta_{1n}z) F^2(z_0, \xi_n)$$

$$+ j\frac{8\pi \omega \rho_1}{r} \sum_{n,n \neq m} \sum_m \frac{\beta_{1m}}{\sqrt{\xi_n \xi_m}} \times \sin(\beta_{1n}z) \cos(\beta_{1m}z) F(z_0, \xi_n) F(z_0, \xi_m)$$

$$\{\cos[(\xi_m - \xi_n)r] + j[\sin][(\xi_m - \xi_n)r]\} \tag{8.30}$$

The first term in the above formula is an imaginary number, indicating that the vertical sound intensity flow of each order normal wave is reactive, but the second term in the above formula is a complex number. Due to the cross interference of multiple order normal waves, there is also active sound intensity flow in the z-axis direction (vertical direction), which will cause the sound energy to operate in the vertical direction at some distance sections, which deepens the understanding of the normal wave sound field.

The active component I_{zA} and reactive component I_{zR} of vertical sound intensity flow are

$$
I_{zA} = \text{Re}(pv_z^*) = \frac{8\pi\omega\rho_1}{r} \sum_{n,n\neq m} \sum_m \frac{\beta_{1m}}{\sqrt{\xi_n\xi_m}} \sin(\beta_{1n}z)\cos(\beta_{1m}z)F(z_0,\xi_n)
$$

$$
\times F(z_0,\xi_n)\sin[(\xi_m - \xi_n)r] \tag{8.31}
$$

$$
I_{zR} = \text{Im}(pv_z^*) = \frac{4\pi\omega\rho_1}{r} \sum_n \frac{\beta_{1m}}{\xi_n} \sin(2\beta_{1n}z)F^2(z_0,\xi_n) + \frac{8\pi\omega\rho_1}{r}
$$

$$
\times \sum_{n,n\neq m} \sum_m \frac{\beta_{1m}}{\sqrt{\xi_n\xi_m}} \sin(\beta_{1n}z)\cos(\beta_{1m}z)F(z_0,\xi_n)F(z_0,\xi_n)\cos[(\xi_m - \xi_n)r]
$$

$$
\tag{8.32}
$$

Although the reactive component of sound intensity flow does not deal with transport energy, it is one of the important characteristics of sound field. From the perspective of signal processing, it still carries important information of sound field. See the following section for quantitative analysis of reactive and active components of horizontal sound intensity flow and vertical sound intensity flow at low frequency.

8.2.3 Low Frequency Normal Wave Sound Intensity Flow Characteristics [10]

From the physical point of view, the participation of active components in acoustic energy transport is the focus of underwater acoustic propagation research. The reactive component does not transport sound energy, but in vector signal processing, the reactive component can also be used for acoustic signal detection, target classification and recognition. In order to serve the latter purpose, it is first necessary to understand the characteristics of reactive power components. Here, the characteristics of the reactive component of low-frequency vertical sound intensity flow are emphatically analyzed, which is shown in Eq. (8.32).

Conditions: sea depth $H = 100$ m, $c_1 = 1480$ m/s, $c_2 = 1550$ m/s, $b = \rho_1/\rho_2 = 1/1.723$, low frequency sound wave with frequency band below 70 Hz, single frequency harmonic sound wave radiated by point source.

According to Eq. (8.22) and the above conditions, the cut-off frequency f_n value is calculated and listed in Table 8.1.

Table 8.1 Cut off frequencies of normal waves in shallow water layer

(n-order)	1	2	3	4	5
f_n/Hz	12.5	37.4	62.3	87.2	112.1

When the acoustic frequency of the sound source is between f_2 and f_3 listed in Table 7.3, only the first-order normal wave exists in shallow water. The accuracy of this film value is very important to the solution of sound field. Taking 50 Hz as an example, the solution is $\beta_{11} = 0.025\ 016\ 348\ 362\ 21$, $\beta_{12} = 0.051\ 099\ 315\ 447\ 41$, $\xi_1 = 0.210\ 790\ 510\ 276\ 06$, $\xi_2 = 0.206\ 027\ 466\ 296\ 68$. The sound pressure amplitude and vibration velocity amplitude distribution of the corresponding first-order normal wave in the vertical direction are shown in Fig. 8.10a and b respectively.

Corresponding to the sound wave with the frequency of 50 Hz, when the receiving vector sensor is located at $z = 36$ m, 38 m and 40 m respectively, if the depth of the sound source changes from water surface $z_0 = 0$ m to seabed $z_0 = 100$ m, the positive and negative signs of the reactive component of the received vertical sound intensity flow are respectively shown in Fig. 8.11a, b and c, and the ordinate in the figure represents the depth of the sound source, The abscissa represents the horizontal distance between the receiving point and the sound source. Black represents the negative sign area and white represents the positive sign area. It can be seen from the figure that when the source depth changes, the sign of the reactive component of the received vertical sound intensity flow changes regularly. For sound waves with frequencies of 40 Hz and 60 Hz, similar results are shown in Fig. 8.11d, e, f and Fig. 8.11g, h, i, respectively. Combining the characteristics of the reactive component of the vertical sound intensity current of the three acoustic waves located between the cut-off frequencies f_2 and f_3 at the receiving point, if $z = 38$ m is selected as the layout position of the receiving vector sensor and $z_0 = 40$ m is set as the "critical depth", the position of the target relative to the critical depth can be judged by the

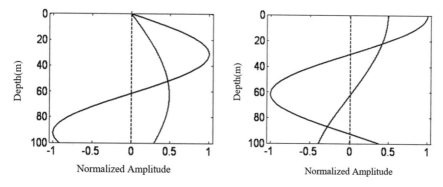

Fig. 8.10 Amplitude distribution of sound pressure and vibration velocity in the vertical direction of the second-order normal wave of the head (50 Hz) **a** sound pressure amplitude distribution of head second-order normal wave **b** vibration velocity amplitude distribution of head second-order normal wave

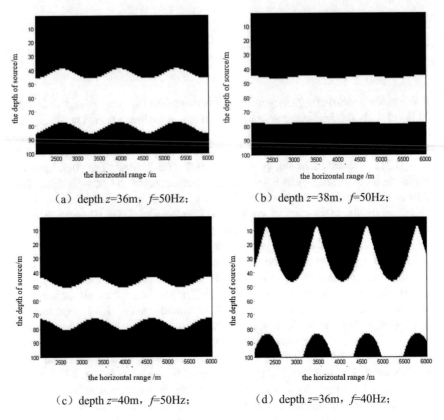

Fig. 8.11 The variation of sign of reactive component of vertical sound intensity flow with sound source depth (black is negative, white is positive, reception depth is z, and sound frequency is f)

positive and negative reactive component of the vertical sound intensity current. In this case, in order to ensure navigation safety, the target cannot navigate at a depth greater than 80 m, and the sea depth is 100 m. In this way, the target can be divided into two categories (surface target and underwater target) according to the relationship between the target navigation depth and the critical depth. In particular, it should be noted that the placement depth of the receiving sensor is not arbitrary. In this example, it can be seen that the placement depth should be 38 m, but the accuracy of the depth is not high.

It can be seen from Sect. 8.2.2 that the sound intensity flow has both active and reactive components in both horizontal and vertical directions. Suppose that the receiving vector sensor is located at $z = 38$ m and the horizontal distance from the sound source is 4 km. When the sound source with frequency of 50 Hz changes within the sea depth range, each component of the received sound intensity flow is shown in Fig. 8.12a. The size of each component in the figure has been normalized to the maximum value of the active component I_{rA} of the horizontal sound intensity

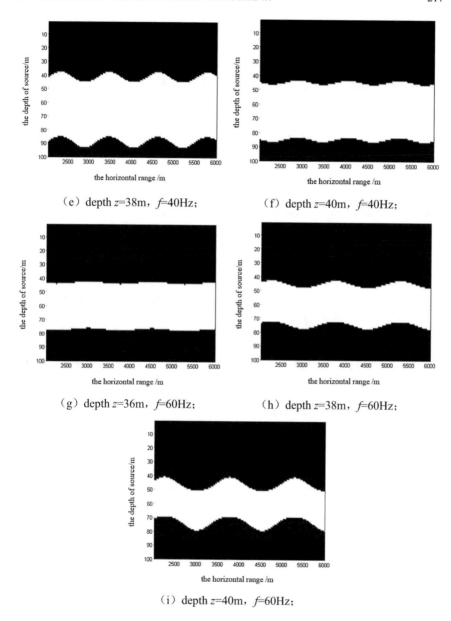

（e）depth z=38m, f=40Hz;　　　　　　　（f）depth z=40m, f=40Hz;

（g）depth z=36m, f=60Hz;　　　　　　　（h）depth z=38m, f=60Hz;

（i）depth z=40m, f=60Hz;

Fig. 8.11 (continued)

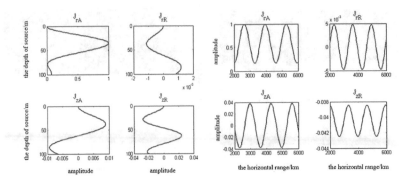

(a) Receiving depth z=38m, horizontal distance from sound source r=4km; (b) Sound source depth $z_0 = 2$m, receiving depth z=38m

Fig. 8.12 The relative amplitude distribution of each component of sound intensity flow (50 Hz)

flow. I_{rR}, I_{zA}, I_{zR} represents the reactive component and active and reactive components of vertical sound intensity flow. Due to the interaction of normal waves, each component of sound intensity flow is finite and has no zero term. When the sound source with frequency of 50 Hz is located at underwater $z_0 = 2$ m, the variation of each component of sound intensity current with distance at the horizontal distance of 2–6 km and the depth of $z = 38$ m is shown in Fig. 8.12b (the attenuation factor according to cylindrical wave has been deducted and normalized according to the maximum value of I_{rA}). The amplitude of I_{rR} and I_{zA} changes periodically, leading to additional changes in signal-to-noise ratio at different distances or depths, which is of guiding significance to analyze the azimuth estimation results of sound intensifier based on I_{rA}. I_{rR} and I_{zA} fluctuate with distance, while I_{zR} fluctuates less with distance under the condition of this example, which again shows that the reactive component of vertical sound intensity flow can be used as the object of signal processing.

8.2.4   The Cross Spectrum Processor

According to the above research results, the cross-spectrum processor block diagram can be obtained, as shown in Fig. 8.13.

The common conjugate spectrum analysis of $p(t)$, $v_z(t)$ was carried out to obtain $P(f)V_z^*(f)$. Where,

$$p(t) \Leftrightarrow P(f)$$
$$v_z(t) \Leftrightarrow V_z(f)$$

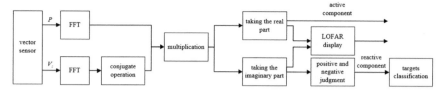

Fig. 8.13 The block diagram of cross spectrum processor P, V_z

The active component is $\text{Re}\left[P(f)V_z^*(f)\right]$ and the reactive component is $\text{Im}\left[P(f)V_z^*(f)\right]$. For the radiation spectrum of the detection target, if $\text{Im}\left[P(f)V_z^*(f)\right]$ is negative, it will be considered that the target is less than the critical depth, otherwise the target is greater than the critical depth. In engineering application, the location of vector sensors should be determined according to the specific environmental conditions of the sea area.

Low frequency sound propagation has attracted more and more attention. In this chapter, the description and basic characteristics of sound intensity flow of point source sound field in Pekeris waveguide are discussed. The total sound intensity flow includes not only the sound intensity flow of each order of normal waves, but also the energy of mutual interference between each order of normal waves. The active component of horizontal sound intensity current changes periodically with the change of distance, which will affect the received signal-to-noise ratio. Although the reactive component of vertical sound intensity flow does not participate in the processing of sound energy, when a single sound vector sensor is properly placed, it can be used to identify the specific depth of the target in signal processing, which is of application value for vector passive sonar and so on.

8.3 Waveguide Invariants β

There is a stable coherent structure in low frequency or very low frequency shallow water sound field. The coherent structure of sound field can be measured by single hydrophone, single vector sensor, vertical hydrophone array, vertical vector array, horizontal hydrophone array or horizontal vector array. The output of the sensor or sensor array is transformed in a variety of time and space, and the coherent structure of the sound field is described in a variety of transformation domains. The transformation depends on the purpose of application and research. There are many characteristic functions describing the coherent structure, such as: the power spectrum $P(r, \omega)$, $P(z, \omega)$ of the received signal; LOFAR diagram $P(f, t)$, $P(z, t)$; Or sound intensity current $I(r, \omega)$, $I(z, \omega)$; $I(t, \omega) \ldots$, etc.

In the LOFAR diagram in Sect. 8.1, replace the time coordinate with $r = \upsilon t$, and υ is the moving speed of the source, and the $P(r, f)$ interferogram or PRW interferogram is obtained. The characteristics of interference fringes in $P(r, \omega)$ diagram are only determined by the environmental conditions of the channel, while those in LOFAR diagram are also related to the motion characteristics of sound source. $P(r, \omega)$ can be predicted theoretically. If $P(r, \omega)$ is measured, the target motion parameters can be estimated according to both. Section 8.1 introduces the interference structure of short-range sound field, and this section introduces the basic concept of interference structure of long-range sound field.

If the sound source emits broadband continuous spectrum signal, the received signal is analyzed by power spectrum to obtain $P(r, \omega)$ diagram. Figure 8.14 shows the sea trial results given in document [11]. Stable interference fringes can be seen from the figure, and it can be observed that the structure of interference fringes is different from other distances in the range of 125–150 km.

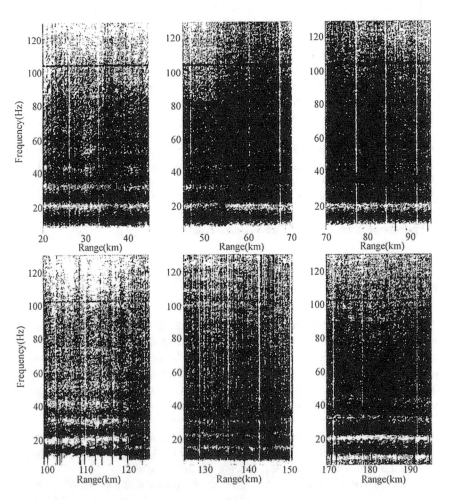

Fig. 8.14 The interference structure $P(r, f)$ of 20–195 km shallow water waveguide with frequency of 5–150 Hz

In 1982, the Russian scholar S. D. Chuprov proposed the concept of channel invariant β, which can be used to describe the slope of interference fringes.

On the power spectrum interferogram $P(r, f)$, the slope η of the interference fringe is

$$\eta = \frac{dr}{df} = \tan \varphi \quad \text{(along the path of an interference fringe)} \tag{8.33}$$

where φ is the angle between the interference fringe and the F-axis.

Chuprov points out that the relationship between waveguide invariant β and interference fringe slope η on $P(r, f)$ diagram is

$$\frac{1}{\beta} = \frac{f}{r}\frac{dr}{df} = \frac{f}{r}\eta = \frac{f}{r}\tan \varphi = \frac{\omega}{r}\frac{dr}{d\omega} \tag{8.34}$$

and

$$\beta = -\frac{r}{\omega}\frac{d\omega}{dr} = \frac{r}{f \tan \varphi} = \frac{d\left(\frac{1}{u}\right)}{d\left(\frac{1}{v}\right)} \tag{8.35}$$

where u is the phase velocity and v is the group velocity. β is the waveguide invariant, and φ is the angle between the interference fringe and the f-axis on the $P(r, f)$ diagram.

Chuprov points out that for layered dielectric waveguides, the value of β is independent of distance at a long distance, and there is $\beta = 1$ in all reverse interface uniform waveguides. In the transition sea area with uneven seabed inclination, the value of β is related to distance and depends on environmental conditions.

The numerical calculation software of sound field can be used to predict the $P(r, \omega)$ degree of sound field interference structure and calculate the β value. The relevant detailed theory is beyond the scope of this tutorial. Please refer to the relevant document [11], which will not be introduced here. This section only introduces the above basic concepts.

According to the LOFAR diagram $P(r, \omega)$ measured by the receiving array, the signal processing technology based on waveguide invariant β can realize long-range passive ranging and velocity measurement of the target, which is an important new sonar technology in development.

Fig. 8.15 The structure
diagram of shallow water
waveguide interference

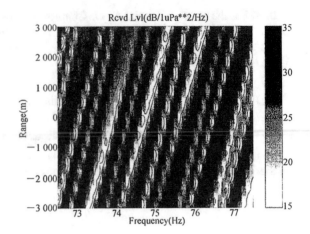

References

1. Hui J, Sun G. Normal mode acoustic intensity flux in Pekeris waveguide and its cross spectra signal processing. Acta Acustica (Chinese version). 2008.
2. Brekhovskikh L. Waves in layered media. Elsevier; 2012.
3. Tolstoy I, Clay CS. Ocean acoustic. McGraw-Hill Book Company; 1977.
4. Wang DZ, Shang EC. Underwater acoustics. Beijing: Science Press; 1984.
5. Boyles CA. Acoustic waveguides: applications to oceanic science. 1984.
6. Yang SE. Principle of underwater acoustic propagation. Harbin: Harbin Engineering University Press; 1994.
7. Shchurov VA. Vector acoustics of the ocean. Vladivostok: Dalnauka; 2006.
8. Hui J, Hu D. Researches on the measurement of distribution image of radiated noise focused beamforming. Acta Acustica (Chinese version). 2007.
9. Gonzalez RC. Digital image processing. Pearson Education India; 2009.
10. Kuperman WA, et al. Ocean acoustic interference phenomena and signal processing. AIP Conference Proceedings; 2001.
11. Yang J, Hui J. Moving target parameter estimation using the spectrum interference of low frequency. J Harbin Inst Technol. 2007.

Chapter 9
Examples in Application

Although this chapter involves a variety of special technologies, it does not focus on the in-depth analysis of special technologies, but as an example in application for learning the basic knowledge of this book.

9.1 The Optimum Operating Frequency of Interrogation Transponder

Response ranging technology is widely used in underwater acoustic navigation and positioning. The interrogator sends out an interrogation sound pulse, and the transponder at distance r answers a reply sound pulse after receiving the signal. The interrogator measures the propagation delay τ of interrogation reply, and the distance r is:

$$r = \frac{c\tau}{2} \tag{9.1}$$

where, c is the sound velocity.

The receiving and transmitting transducers of transponder and interrogator are horizontal and non-directional with wide vertical directivity. To simplify the discussion, it is assumed that the transducer is three-dimensional directionless.

Assuming that the interrogator and transponder are stationary, the background interference is marine environmental noise.

The so-called optimal operating frequency refers to the operating frequency that saves the power most at a given operating distance.

The passive sound equation applied to the above problem is:

$$SL - TL - NL \geq DT \tag{9.2}$$

© Harbin Engineering University Press 2022
J. Hui and X. Sheng, *Underwater Acoustic Channel*,
https://doi.org/10.1007/978-981-19-0774-6_9

where, *TL* is:

$$TL = 20 \log r + 60 + \alpha r \tag{9.3}$$

$$\alpha = 0.036 f^{3/2} \quad (\text{dB/km}) \tag{9.4}$$

The above equation is an empirical formula obtained from experiments.
The interference level *NL* is:

$$NL = 63 - 20 \log f + 10 \log B \quad (f > 1\text{kHz}) \tag{9.5}$$

where, the unit of f is kHz and B refers to bandwidth. Equation (9.5) is the approximate fitting of the experimental curve in Fig. 2.9.

At the maximum action distance, Eq. (9.2) takes the equal sign.

Equation (9.2) derivatives f. When the derivative is equal to zero, the equation followed by the optimal operating frequency f_0 is obtained:

$$\frac{20}{f_0} - 0.036 \times \frac{3}{2} \times f_0^{\frac{1}{2}} r = 0$$

$$f_0 = \frac{50}{r^{2/3}} \tag{9.6}$$

The unit of f in the above equation is kHz, and the unit of r is km.

If the designed response action distance is $r = 8\,\text{km}$, the optimal working frequency $f_0 = 12.5\,\text{kHz}$ is obtained from the above formula.

The above equation indicates that the greater the required action distance is, the lower the working frequency will be.

9.2 Passive Sonar Range Estimation

Passive sonar is the main battle sonar of submarines. Modern submarines are equipped with bow array sonar, side array sonar and towed linear array sonar. They have their own advantages and perform their own duties, among which the towed linear array sonar has the farthest distance. Discussing the optimization technology of each sonar is beyond the purpose of this book. This section only summarizes the application of sonar equation and the selection of sonar parameters.

The passive sonar equation is:

$$SL - TL - NL + G_s + G_t \geq DT \tag{9.7}$$

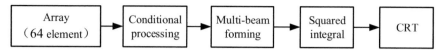

Fig. 9.1 The schematic block diagram of towed linear array sonar

Compared with Eq. (1.25), G_s and G_t replace the directivity index DI. G_s is the spatial gain of the array and G_t is the time gain of post-processing. In fact, the spatial filtering gain of the array must be described by the spatiotemporal correlation matrix of the array. Equations (1.25) and (9.7) are approximate, but the latter is more appropriate.

$(SL$-TL-$NL)$ is the received signal to clutter ratio of each array element of the array. The left of Eq. (9.7) is the input signal-to-noise ratio of the display (see Fig. 9.1) . DT is the minimum input signal-to-noise ratio required when the display terminal reaches the confidence limit.

According to Eq. (1.25), the quality factor GL of passive sonar is:

$$GL = SL - NL + G_s + G_t - DT \tag{9.8}$$

The greater the GL, the farther the action distance, and the lower the optimal working frequency. We can refer to Fig. 1.6. The array of towed linear array sonar is about 1000 m behind the stern of the boat, so it is less disturbed by the radiated noise of the boat than the hull sonar; The scale of the towed array can be much larger than that of the Hull Sonar, and the larger array aperture has higher spatial gain. In short, the quality factor of towed array sonar is greater than that of boat shell sonar, with the farthest operating distance and the lowest operating frequency band. Therefore, this section only discusses the operating range estimation and sonar parameter selection of towed array sonar.

The drag array is a uniform linear array with equal spacing. Let the array length be $L = 189$ m. The array element spacing is 3m, and there are 64 array elements in total. The center frequency f_0 is 250 Hz and the working frequency band is 125 ~ 500 Hz.

$$f_0 = \sqrt{f_H f_L} \tag{9.9}$$

where f_H is the upper cut-off frequency and f_L is the lower cut-off frequency.

This frequency band is the sound frequency band radiated by the target mechanical vibration, and the frequency band with the strongest continuous spectrum of radiated sound. The target sound source level of this frequency band is higher than that of other frequency bands. When the operating frequency band is higher than 500 Hz, the beam pattern of the array will have sub maximum grid lobes, resulting in direction finding ambiguity and increasing the intensity of background interference. In the frequency band below 100 Hz, the flow noise is very strong, so the value should be greater than this value.

Background interference sources of towed array sonar include:

- Flow noise: the pressure fluctuation noise in the turbulent boundary layer when the array is moving is called flow noise.
- Streamer vibration interference.
- Towed craft radiated noise interference.
- Marine environmental noise.

If special technology is adopted to suppress the current noise and the boat is towed at low speed, the current noise interference can be lower than that of marine environment. The array technology to suppress flow noise is the key technology.

A damping section is inserted between the streamer and the array to effectively suppress the vibration interference.

Towed ship radiated noise is an important interference and a kind of coherent source interference. It produces a detection blind area at the end of the array. The blind area can be appropriately reduced by using adaptive tugboat interference cancellation technology.

To sum up, marine environmental noise is the main interference background. It is isotropic interference, and its spatial correlation radius is half wavelength. Therefore, the output interference of each element of the half wave spacing array is uncorrelated. The spatial gain G_s of the array is:

$$G_s = 10 \log N \cong 18 \, \text{dB} \qquad (9.10)$$

$N = 64$ (for $f = 250$ Hz), for the low-end frequency component, the G_s value should be reduced by 3 dB.

The principle block diagram of towed linear array sonar is shown in Fig. 9.1.

The square integrator is a post processor, and its anti-interference gain G_t is:

$$G_t = 5 \log BT = 23 \, \text{dB} \qquad (9.11)$$

where, B is the bandwidth, $B = f_H - f_L$. T is the integration time length, and take $T = 50$ s.

Just to simplify the calculation, assuming that the power spectrum shape of the continuous spectral acoustic radiation of the target is the same as that of the ambient noise interference, SL and NL in the sonar equation can be calculated by spectral level.

For conventional submarines:

$$SL = 115 \, \text{dB} \quad (500 \, \text{Hz}, 1 \, \text{Hz})$$

Referring to Fig. 2.9, the level 3 sea state interference spectrum is:

$$NL = 68 \, \text{dB} \quad (500 \, \text{Hz}, 1 \, \text{Hz})$$

Substitute into the sonar Eq. (9.7) and obtain: (Take $DT = 3$ dB)

$$GL = TL \leq 85 \text{ dB} \tag{9.12}$$

Transmission loss TL is:

$$TL = 60 + 15 \log r + \alpha r = 85 \text{ dB} \tag{9.13}$$

The value of α is very small and αr can be ignored. According to the solution of Eq. (9.13), the action distance is about 80 km.

Low frequency passive sonar is a long-range detection sonar, which must be carefully designed and combined. If the system loses 3 dB in any link, the action distance will be doubled.

9.3 Fundamentals of Three Element Array Passive Ranging Sonar

In order to ensure concealment, submarines generally do not use active sonar ranging, so passive ranging technology must be employed. Modern passive ranging technologies include:

- Three element array passive ranging.
- Matched field passive ranging.
- Time reversal mirror passive ranging.

The latter two are developing passive ranging technologies, which can realize long-range passive ranging. Three element array passive ranging is a mature technology with high ranging accuracy in medium and short range. This section introduces its principle, especially focusing on the problems related to its matching with acoustic channel.

Three primitives, or three flat arrays, are installed on the side of the submarine, with one group on each side. A typical ternary array is a symmetrical array with equal spacing, as shown in Fig. 9.2.

The array element spacing is d, and the cylindrical coordinates of target s are (r, θ). Only plane problems are considered. The distance r_i from the target to the three array elements is:

$$r_1^2 = r^2 + d^2 - 2dr \cos \theta \tag{9.14}$$

$$r_2 = r \tag{9.15}$$

$$r_3^2 = r^2 + d^2 + 2dr \cos \theta \tag{9.16}$$

Fig. 9.2 schematic diagram
of three element array
passive ranging

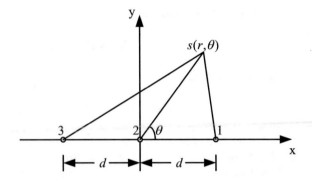

θ is the target azimuth, and the bow direction is $0°$, r is the distance from the target to the central array element.

Subtract Eq. (9.14) from Eq. (9.16) to obtain:

$$4dr \cos\theta = (r_3 - r_1)(r_3 + r_1) \tag{9.17}$$

At a distance, approximately:

$$r_3 + r_1 \approx 2r$$

Substitute into Eq. (9.17) to obtain:

$$\theta = \cos^{-1}\frac{r_3 - r_1}{2d} = \cos^{-1}\frac{c\tau_{31}}{2d} \tag{9.18}$$

In the above formula, c is the sound velocity, and τ_{31} is the delay difference between the target acoustic signal reaching array element 3 and array element 1. τ_{31} is estimated by cross correlator, and the target azimuth can be calculated according to the above equation, which is actually the result of plane wave approximation.

Add $(r^2 - 2rr_1)$ on both sides of Eq. (9.14) to obtain:

$$(r_1 - r_2)^2 = 2r^2 + d^2 - 2rr_1 - 2dr \cos\theta \tag{9.19}$$

Add $(r^2 - 2rr_3)$ on both sides of Eq. (9.16) to obtain:

$$(r_3 - r_2)^2 = 2r^2 + d^2 - 2rr_3 + 2dr \cos\theta \tag{9.20}$$

Subtract Eq. (9.19) from Eq. (9.20) to obtain:

$$[(r_3 - r_2) + (r_1 - r_2)][(r_3 - r_2) - (r_1 - r_2)] = 2r(r_1 - r_3) + 4dr \cos\theta$$

Substitute Eq. (9.17) into the above equation to obtain:

$$[(r_3 - r_2) + (r_1 - r_2)] = (r_3 + r_1) - 2r$$

$$c[\tau_{32} + \tau_{12}] = (r_3 + r_1) - 2r \tag{9.21}$$

$$r_3 + r_1 = \sqrt{r^2 + d^2 + 2dr\cos\theta} + \sqrt{r^2 + d^2 - 2dr\cos\theta}$$

When $r \gg d$, the above equation is approximately:

$$r_3 + r_1 \cong \sqrt{r^2 + 2dr\cos\theta} + \sqrt{r^2 - 2dr\cos\theta}$$

$$\cong 2r + \frac{d^2\cos^2\theta}{r} \tag{9.22}$$

By substituting this into Eq. (9.21), we obtain:

$$c(\tau_{12} + \tau_{32}) \cong \frac{d^2\cos^2\theta}{r}$$

$$\frac{1}{r} \cong \frac{c(\tau_{12} + \tau_{23})}{d^2\cos^2\theta} \tag{9.23}$$

The cross correlator is used to estimate τ_{12} and τ_{23}, $\cos\theta$ is calculated by Eq. (9.18), and then the target distance r can be calculated by Eq. (9.23).

The greater the length $2d$ of the array, the higher the ranging accuracy, but the length of the array shall not exceed the spatial correlation radius of the acoustic channel. It has been pointed out in Sect. 3.7 that when the array length is greater than the spatial correlation radius, the correlation number will decrease and the delay difference estimation error will increase. The experimental data provided in Table 3.4 and Fig. 3.19 show that the spatial correlation radius of CW signal is about 40 ~ 47 m (1 ~ 9 kHz experimental results). Therefore, the length 2D of the ternary array should be 24 ~ 45 m.

The spatial correlation radius of single frequency signal is larger than that of broadband signal. The target radiated noise is broadband. It has been pointed out in Sect. 3.3 that the main peak of ambiguity function of broadband signal is sharp and has better time delay difference estimation accuracy. Therefore, multi band integrated time delay estimation should be used in passive ranging sonar, which can reduce the benefits of strong spatial correlation of narrow band signals and high accuracy of broadband time delay estimation.

In Sect. 5.7, it is pointed out that the target motion will produce Doppler cross-correlation loss, and the calculation formulas of allowable cross correlator integration time length and post integration time length are given. Doppler cross-correlation loss must be considered in the design of passive ranging cross-correlator.

9.4 Passive Detection Principle of Single Vector Sensor

Single vector sensor passive detection technology has a wide range of applications. Such as radio passive directional underwater acoustic buoy, mine acoustic fuze, underwater information network detection node, coastal early warning acoustic detector, marine environmental noise and biological noise monitoring, etc. It is favored because of its simplicity and low power consumption. In particular, it can use small-size sensors to realize high-precision direction finding and remote detection of broadband low-frequency sound sources. This section only gives a basic introduction, focusing on the principle of vector signal DF method.

This section is slightly expanded on the basis of Sect. 1.5.

The average sound intensity generator in Fig. 1.6 is the time domain processing method of vector signal processing. In the isotropic noise background, the vibration velocity and sound pressure are uncorrelated, and the expectation of the average sound intensity flow is 0. In fact, the interference sound intensity stream output by the average sound intensifier jitters around the value of 0. The sound pressure and vibration velocity of remote target signal are completely correlated. This makes the anti-interference effect of the average sound intensity device equivalent to that of an independent primitive cross correlator.

Sound wave is a longitudinal wave. Under the condition of plane wave, the direction of vibration velocity or sound intensity flow is the direction of sound source. The far field of any complex traveling wave sound field can be approximated as a plane wave. Ohm's law of acoustics is established, and the target orientation is the direction of sound intensity flow. It can be seen from Fig. 9.3 that the target azimuth θ is the arctangent of the ratio of two sound intensity current components. We have:

$$\theta = tg^{-1} \frac{\overline{pv_y}}{\overline{pv_x}} \tag{9.24}$$

In the above equation, θ is the orientation of the target in the sensor coordinate system, and the x axis is $0°$. $\overline{pv_x}$ and $\overline{pv_y}$ are two orthogonal average sound intensity current components. The above equation holds under the plane wave approximation. In the ocean acoustic channel, the error of Eq. (8.24) becomes larger for very low

Fig. 9.3 Direction finding principle of average sound intensity flow

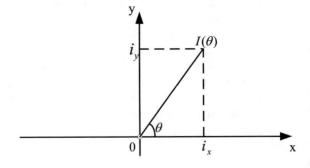

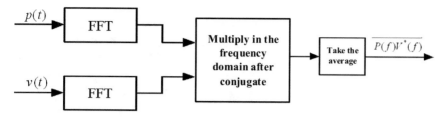

Fig. 9.4 Sound intensifier

frequency acoustic waves in short range, because there is a frequency dependent phase difference between sound pressure and vibration velocity.

The frequency domain processing corresponding to Eq. (9.24) is called complex sound intensifier, and its block diagram is shown in the Fig. 9.4.

The complex sound intensifier is the cross spectrum processor of sound pressure and vibration velocity, and its output is cross spectrum $\overline{P(f)V_x^*(f)}$ or $\overline{P(f)V_y^*(f)}$. Target azimuth $\theta(f)$ is:

$$\theta(f) = tg^{-1} \frac{\overline{P(f)V_y^*(f)}}{\overline{P(f)V_x^*(f)}} \tag{9.25}$$

For continuous spectrum acoustic radiation target, azimuth θ is:

$$\hat{\theta} = <\theta(f)>_f \tag{9.26}$$

$<\cdot>_f$ represents the set average in the frequency domain.

The target radiated noise is not white, so Eq. (9.25) has higher DF accuracy than Eq. (9.24).

Figure 9.5 shows the sea trial results of surface ship detected by single vector sensor. The horizontal coordinate of waterfall map is azimuth and the vertical coordinate is time. It is an azimuth process. The maximum distance of the target in the map is 13.6 km. The target ship is about 6000 tons and the speed is 9 knots. Sea state level is 3 with negative gradient hydrology and sea depth of 55 m. The vector sensor is suspended 25 m deep on the side. Due to the lack of vibration reduction measures, the disturbance caused by ship turbulence is the major one. Nevertheless, the detection distance of single vector sensor is still satisfactory.

9.5 Moving Target Detector

Moving target detector is an effective detection technology to detect moving targets in reverberation background. It is used to resist reverberation in active sonar. Active torpedo acoustic homing system is an unmanned operating system. The required false

Fig. 9.5 The sea trial results
of single vector sensor
detection

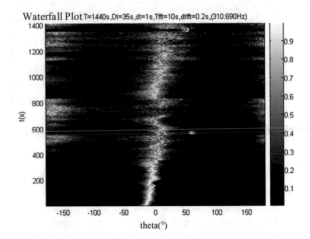

alarm probability is very low, and false alarm is almost not allowed. Reverberation is the most important interference source causing false alarm. Therefore, this section takes this as an example to discuss moving target detection technology and pay attention to the matching between the system and reverberation channel.

Firstly, based on Chap. 6, the reverberation channel characteristics involved in Torpedo active acoustic homing are analyzed.

It is pointed out in Sect. 6.2 that the power spectrum of reverberation is roughly the same as the shape of transmitted signal. Two important characteristics of reverberation must be paid attention to:

- Doppler shift

If the Doppler shift caused by the motion of homing platform (torpedo) has been compensated, the Doppler shift of seabed reverberation is 0; The Doppler shift of volume reverberation is caused by the velocity of seawater, usually no more than 2 knots; The Doppler shift of sea surface reverberation is slightly larger, depending on the traveling speed of swell. The Doppler frequency shift of the target echo is large, unless the sound axis of the beam is perpendicular to the direction of the target speed, the echo Doppler is zero. Therefore, the low Doppler target is not necessarily a low-speed target. Moving target detection technology relies on the Doppler difference between target echo and reverberation to distinguish echo spectrum and reverberation spectrum.

- Doppler frequency spread of reverberation spectrum

The reverberation spectrum, called Doppler spread, is slightly wider than the emission spectrum. The broadening according to size from large to small is: seabed reverberation, volume reverberation and sea surface reverberation. The Doppler broadening of the latter is about 1.75 knots. If the operating frequency is 30 kHz, the Doppler broadening is equivalent to 4.7 Hz.

The Doppler shift during two-way transmission is:

$$\Delta f = 0.7 f_0 v \cos \theta \ (\text{Hz}) \tag{9.27}$$

f_0 is the center frequency of the long CW pulse transmission signal, in kHz, v is the target speed in knots, and θ is the angle between the sound propagation direction and the speed direction.

According to the above equation, the Doppler frequency shift is 4.7 Hz. Only when the Doppler velocity of the target is greater than 1.75 knots can the target echo spectrum be separated from the reverberation spectrum with Doppler spread. Therefore, moving target detector can only effectively detect targets whose Doppler shift is greater than Doppler spread.

In order to separate the echo spectrum from the reverberation spectrum, the transmitted signal spectrum must be narrow enough, or the ambiguity function of the transmitted signal should have high frequency resolution. The frequency resolution of CW pulse is the reciprocal of pulse width T, i.e. $\frac{1}{T} \le 4.7$ Hz ($f_0 = 30$ kHz), which requires pulse width $T > 250$ ms.

The moving target detection system emits long CW pulse, and its spectral width should be less than the Doppler broadening of reverberation.

The Doppler shift and Doppler spread of reverberation are different for beams pointing in different directions; The Doppler shift and spread of the three types of reverberation are also different. These three kinds of reverberation arrive at different times, so the frequency shift and spread of the same beam output are also different. This shows that only adaptive anti reverberation processing can match the above time-varying and space-varying reverberation, so as to achieve better anti reverberation effect.

Let's estimate the average reverberation level.

What is the reverberation limit distance of the acoustic homing system?

The so-called reverberation limit distance refers to the critical distance where the reverberation level is equal to the self-noise level. Within the reverberation limit distance, reverberation is the main interference.

The size of the self-guided array is about 400 mm, the CW pulse working frequency is 30 kHz, the pulse width is 250 ms, the emission sound source level $SL = 225$ dB, and the self-noise spectrum level $NL_0 = 70$ dB (30 kHz, 1 Hz, re:1 μ Pa) where the array is located. Horizontal beam width $\theta_H = 7.5°$, vertical beam width $\theta_r = 30°$. The target strength is assumed to be.

According to Eq. (6.1), the average reverberation intensity is:

$$RL_{s \cdot v} = SL - 2TL + S_{s,v} + 10 \log A, \ V$$

If $r = 2$ km, there is:

$$TL = 60 + 20 \log r + \alpha r$$

If $f_0 = 30$ kHz, $\alpha \approx 5.8$ dB/km.

$$TL = 60 + 6 + 11.6 = 77.6 \tag{9.28}$$

Substituting into Eq. (6.1), the average reverberation level at 2 km is obtained as follows: (Take $S_v = -70\,\text{dB}$)

$$RL_v = 225 - 155.2 - 70 + 10\log V \tag{9.29}$$

$$V \cong (r\theta_H)(r\theta_V)\frac{CT}{2}$$

$$= 4 \times 10^6 \times 0.125 \times 0.6 \times \frac{1500 \times 0.25}{2}$$

$$= 5.63 \times 10^6$$

Substituting into Eq. (9.29), we have:

$$RL_v = 67\,\text{dB} \tag{9.30}$$

The above equation estimates that the volume reverberation level at 2 km is lower than the self-noise spectral level, so the volume reverberation is not important.

The scattering coefficient S_s of sea surface reverberation is 10°, and the value of $f_0 = 30\,\text{kHz}$ is $S_s = -30\,\text{dB}$.

For the S_s value of seabed reverberation, take the medium to large value (sweep angle is 10°) in Fig. 6.4 and obtain $S_s = -28$ dB. At 2 km, the reverberation area A is:

$$A \approx (r\theta_H)\frac{CT}{2} = 2 \times 10^3 \times 0.125 \times 187.5$$

$$= 4.7 \times 10^4$$

Substitute into Eq. (6.1) and obtain:

$$RL_s = 225 - 155.2 - 30 + 47$$

$$= 86.8\,\text{dB} \quad (Sea\ surface\ reverberation) \tag{9.31}$$

$$RL_s = 88.8\,\text{dB} \quad (Seabed\ reverberation) \tag{9.32}$$

The moving target detector is adopted. Within the signal bandwidth of 4.7 Hz (T = 0.25 s), the self-noise level NL is:

$$NL = NL_0 + 10\log B$$

$$= 70 + 10\log 4.7$$

$$= 76\,\text{dB} \tag{9.33}$$

According to Eqs. (9.30) ~ (9.33), the limited distance of interface reverberation is greater than 2 km, and the mixing noise ratio is about 10 dB or more.

Then we estimate the signal mixing ratio. The echo intensity EL received at 2 km is:

$$EL = SL - 2TL + T_s \tag{9.34}$$

The signal mixing ratio $\left(\frac{S}{N}\right)_R$ obtained by subtracting Eq. (6.1) from Eq. (9.34) is:

$$
\begin{aligned}
\left(\frac{S}{N}\right)_R &= T_s - 10 \log S_s - 10 \log A \\
&= 15 + 30 - 47 \\
&= -2\,\text{dB} \quad (Sea\ surface\ reverberation)
\end{aligned} \tag{9.35}
$$

$$\left(\frac{S}{N}\right)_R = -4\,\text{dB} \quad (Seabed\ reverberation) \tag{9.36}$$

The above estimation shows that the signal-to-mixing ratio is very low at 2 km. Only when the echo spectrum and reverberation spectrum are separated due to the Doppler frequency shift difference, the reverberation interference will not affect the echo detection. Moving target detector can achieve this goal.

The echo intensity EL at 2 km is obtained from Eq. (9.34):

$$
\begin{aligned}
EL &= 225 - 155.2 + 15 \\
&= 84.8\,\text{dB}
\end{aligned} \tag{9.37}
$$

Equation (9.33) gives the noise interference level, so the signal-to-noise ratio $\left(\frac{S}{N}\right)_n$ is:

$$\left(\frac{S}{N}\right)_n = 84.8 - 76 = 8.8\,\text{dB} \tag{9.38}$$

The above equation shows that a carefully designed moving target detection system may detect targets with target strength $T_s \geq 15\,\text{dB}$ at 2 km.

The principle block diagram of moving target detection system is shown in Fig. 9.6.

The time base controller controls the phase control signal source to generate the phase control transmission signal of CW pulse. The echo intensity of a long CW pulse fluctuates greatly, so the multi frequency system is usually adopted, and the adjacent frequency interval should be greater than the coherent bandwidth of the channel.

When the output signal of the receiving array is amplified and pre filtered for A/D conversion, it must ensure a large enough dynamic range. The multi beam former forms a narrow beam, which has a small reverberation area and is used to

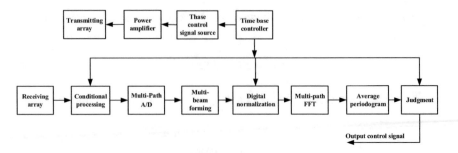

Fig. 9.6 The diagram of moving target detection system

suppress reverberation. After digital normalization, the reverberation after stabilization is output. FFT and average periodogram processing are the core of moving target detection. The sample length of FFT processing matches the signal pulse width T, and the frequency resolution of FFT is 1/T. FFT processing separates the reverberation spectrum from the moving target echo spectrum. Compared with the transmitted signal spectrum, the reverberation spectrum widens 1.75 knots of Doppler, equivalent to 4.7 Hz. For low Doppler target, the reverberation coincides with the echo spectrum, and the system cannot detect the echo. When the Doppler velocity of the target is greater than the threshold, the system detects the target in the noise background, and reverberation does not become the main interference.

A decider automatically learns and stores the reverberation spectrum frequency and suppresses the signal from the transmitting string to the receiving array. The decider automatically determines whether there is a target and outputs the control signal to the homing navigation system.

A moving target detection system is a complex high technology. This section only attaches importance to the principle introduction, focusing on the matching between the moving target detection system and the reverberation channel, and the reverberation level estimation of the active sonar.

References

1. Chen G, Xu JH. Modified coherent matching for shallow sea channel. ACTA ACUSTICA (Chinese version); 1983.
2. Guthrie AN. Long-range low frequnecy C W propagation in the ocean: Antigua-Newfoundland. J.A.S.A; 1974(56):1.
3. Middleton D. A statistical theory of reverberation and similar first-order scattered fields. IEEE Trans Inf Theory. 1967;17:13.
4. Ol'shevskii VV, Middleton D. Statistical methods in sonar[M]. New York: Consultants Bureau; 1978.

5. Fisher RA. Optical phase conjugation. New York: Academic Press; 1983.
6. Nikoonahad TLP. Real time ultrasonic phase conjugation, IEEE. Ultrasonic Symp Proc.; 1989: 677–679
7. Hui JY, Hui J. Fundamentals of vector acoustic signal processing[M]. Beijing: National Defense Industry Press; 2009.

Printed in the United States
by Baker & Taylor Publisher Services